IN DEFENSE OF SCIENTISM

AN INSIDER'S VIEW OF SCIENCE

By

Byron K. Jennings

Vancouver, BC, Canada

ISBN: 978-0-9940589-0-4 (epub), 978-0-9940589-2-8 (epub), 978-0-9940589-3-5 (epub), 978-0-9940589-4-2 (epub), 978-0-9940589-1-1 (paperback)

Only the oxen are consistent.

Old Hungarian Proverb

FORWARD

First launched in 2005 for the International World Year of Physics, the *Quantum Diaries* blog offers a personal glimpse into the daily lives of particle physicists. Through their stories, videos, photos and musings, physicists from universities and laboratories in North America, Asia and Europe share their lives with readers from all backgrounds and in all regions of the world.

With posts about their experiences, impressions, triumphs and disappointments, the *Quantum Diaries* bloggers offer a sense of how scientists think, behave and accomplish their goals—from the triumph of discovering the Higgs boson to the ongoing hunt for dark matter.

Organized by the InterAction Collaboration, which seeks to support the science of particle physics while enabling peaceful collaboration across borders, *Quantum Diaries* is not just about physics; it's about being a physicist. The diarists write about their families, hobbies and interests, as well as their latest research findings and challenges.

Since 2011, Byron Jennings has been one of a dozen active bloggers on the site. His essays are typical of the eclectic mix of *Quantum Diaries* posts, offering insight not only into physics, but also into the mind of a physicist. Byron often philosophizes about the nature of science, revealing a witty and keen mind eager to understand all aspects of the world around him. As he himself wrote in an early essay:

> *The novice and expert view the field differently. Whether it is chess or biology, the novice tends to view the field as a catalogue of facts (an idea that goes back at least to Aristotle), and the important point is what are these facts. ... The expert on the other hand sees patterns, relationships and organization but has no catalogue of true statements. ... For the expert, it is the patterns and relationships that are important.*

As the editor of *Quantum Diaries*, it is my hope that this compilation of essays will offer a glimpse of the patterns, relationships and organization that Byron, as an expert in physics, sees around him every day. Byron's posts have been some of the most viewed on *Quantum Dia-*

ries, and often trigger in-depth conversations in the comments section. I hope that this compilation will further the conversation, encouraging readers to think deeply about the intersection of science and philosophy.

—Kelen Tuttle, Former Editor, *Quantum Diaries*

PREFACE

While this series of essays is normally referred to as philosophy of science, I prefer to think of them as the ruminations of an old timer who has been there and done that. They are one scientist's attempt to understand science on its own terms, from the inside, using the techniques of science itself. I present a model of how science works and discuss the various subtleties and misunderstandings that plague science. This work has arisen from a lifelong interest in science and how it works. As for a scientist and science administrator trying to do something that is nominally philosophy, I note the philosopher C.I. Lewis's contention that everyone both can and must be their own philosopher.

This series of essays was first posted on the QUANTUM DIARIES website run by a consortium of particle physics laboratories including my employer, TRIUMF. However, the opinions expressed are my own. Or at least I held them when I wrote any particular essay. The title of the book is from the title of one of the essays, but the entire book defends the idea that knowledge is only obtained through the proper application of the scientific method. Although the essays were initially published individually, they were designed to give a coherent whole that describes the many facets of how science works. The Prologue ties them into the historical developments in philosophy. While the essays each present a separate thesis and can be read individually, they are best read sequentially since the later essays use concepts discussed earlier. To give a context to the various dates mentioned in the book, a brief timeline for science is given in Appendix A. The essays, in some cases, have editing and other changes from those that appeared in Quantum Diaries.

The essays were edited by J. Gagné, M. Baluk, K. Shauer, or J. Pitcher before they were posted on *Quantum Diaries* but the *ideocentricities* are my own. The final collection was proof read by my wife who found an embarrassing large number of (she says risible) errors. The model boat on the cover was drawn by C. Eakins and the mathematical model (equations) is by the TRIUMF Theory Department. The cover design is by C. Zaworski. I would also like to thank T. Meyer for his help and encouragement. My employer, TRIUMF, is thanked for supporting, or at least tolerating, my foray into philosophy.

TABLE OF CONTENTS

1. PROLOGUE: PHILOSOPHY IS NOT ALWAYS USELESS

This prologue locates my view of the scientific method within the landscape of various philosophical traditions and also ties it into my current interest of project management. As strange as it may seem, this triumvirate of the scientific method, philosophy and management meet in the philosophic tradition known as pragmatism and in the work of W. Edwards Deming (1900 – 1993), a scientist and management guru who was strongly influenced by the pragmatic philosopher C.I. Lewis (1883 – 1964), and who in turn strongly influenced business practices. And I do mean strongly in both cases. The thesis of this essay is that Lewis, the pragmatic philosopher, has had influence in two directions: in business practice and in the philosophy of science. Surprisingly, my views on the scientific method are very much in this pragmatic tradition and not crackpot[1].

The pragmatic movement was started by Charles S. Peirce (1839 – 1914) and further developed by Williams James (1842 – 1910) and John Dewey (1859 – 1952). The basic idea of philosophic pragmatism is given by Peirce in his pragmatic maxim as: *To ascertain the meaning of an intellectual conception one should consider what practical consequences might result from the truth of that conception—and the sum of these consequences constitute the entire meaning of the conception.* Another aspect of the pragmatic approach to philosophic questions was that the scientific method was taken as given with no need for justification from the outside, i.e. the scientific method was used as the definition of knowledge.

How does this differ from the workaday approach to defining knowledge? Traditionally, going back at least to Plato (428/427 or 424/423 BCE – 348/347 BCE) knowledge has been defined as:

- **Knowledge**: justified true belief (Definition 1).

This leaves open the question of how belief is justified (see below) and since no justification is ever 100 per cent certain, we can never

---

[1] Although some people think pragmatism is crackpot.

be sure the belief is true[1]. These considerations are a definite problem and have provided plenty of grist for the philosophical mill for the last two-and-one-half millennia.

A second definition of knowledge predates this and is associated with Protagoras (c. 490 B.C. – c. 420 B.C.) and the sophists:

- **Knowledge:** what you can convince people is true (Definition 2).

Essentially, the argument is that since we cannot know that a belief is true with 100% certainty; what is important is what we can convince people of. This same basic idea shows up in the work of modern philosophers of science with the idea that scientific belief is basically a social phenomenon and what is important is what the community convinces itself is true. This was part of Thomas Kuhn's (1922 – 1996) thesis.

While we cannot know what is true, we can know what is useful. Following the lead of scientists, the pragmatists effectively defined knowledge as:

- **Knowledge:** information that helps predict the future with the possibility of modifying it (Definition 3)

If we take predicting and modifying the future as the practical consequence of information, this definition of knowledge is consistent with the pragmatic maxim. Classical physics is not knowledge by the strict application of Definition 1 since it is not completely true; however it is knowledge by Definition 3 since it helps us predict and modify the future. The modify clause is included in the definition since the pragmatists insisted on that aspect of knowledge. For example, C.I. Lewis said that without the ability to act there is no knowledge. Technology can be thought of as the act part of scientific knowledge; i.e. using information to control, at least, some aspects of the future.

---

[1] International Standards Organization (ISO) 9000 has: *justified belief and having a high certainty to be true.*

2

The scientific method, with its emphasis on the ability to make successful predictions, is built on Definition 3. Or alternately, Definition 3 can be considered a consequence of the scientific method applied globally. The belief in the global applicability of the scientific method is called scientism. Scientism can also be defined as the view that the empirical sciences constitute the sole source of authoritative learning. Needless to say, it is Definition 3 that I prefer, although Definition 1 is rather similar if we follow scientism and insist that the scientific method is the only way to justify belief (the *true* is still a problem). I might add that Definition 2, if the person you are trying to convince believes in the scientific method, also reduces to Definition 3. But it is only Definition 3 that leads to science without any additional assumptions.

The third definition of knowledge given above does not correspond to what many people think of as knowledge so Dewy suggested using the term *warranted assertions* rather than knowledge: the validity of the standard model is a warranted assertion. Fortunately, this terminology never caught on. In contrast, James's pragmatic idea of *truth's cash value*, derided at the time, has caught on. In a recent book HOW TO MEASURE ANYTHING, on risk management, Douglas W. Hubbard spends a lot of space on what is essentially the cash value of information. In business, that is what is important. The pragmatists were, perhaps, just a bit ahead of their time. Hubbard, whether he knows it or not, is a pragmatist. (Hubbard has confirmed that he does indeed know it and that he is a fan of the pragmatic philosophers.)

Dewey coined the term *instrumentalism* to describe the pragmatic approach. As explained by Louis Menand (b. 1952), the approach can be summarized as[1]:

> *An idea or a belief is like a hand, an instrument for coping.*
> *A belief has no more metaphysical status than a fork. When*
> *your fork proves inadequate to the task of eating soup, it*
> *makes little sense to argue about whether there is something*
> *inherent in the nature of forks or something inherent in the*

---

[1] The Metaphysical Club – A Story of Ideas in America, Farrar, Straus and Giroux, New York. See pg.361

nature of soup that accounts for the failure. You just reach for a spoon.

However, most pragmatists did not consider themselves to be instrumentalists but rather used the pragmatic definition of knowledge to define what is meant by real.

Now I turn to C.I. Lewis. He is alternately regarded as the last of the classical pragmatists or the first of the neo-pragmatists. He was quite influential in his day as a professor at Harvard from 1920 to his retirement in 1953. In particular, his 1929 book MIND AND THE WORLD ORDER had a big influence on epistemology and surprisingly on International Standards Organization (ISO) management standards. One can see many of the ideas developed by Kuhn already present in the work of C.I. Lewis[1], for example, the term *paradigm* and the role of theory in interpreting observation. Or as Deming, influenced by Lewis, expressed it: *There is no knowledge without theory.* As a theorist, I like that. At the time, this was quite radical. The logical positivists took the opposite tack and tried to eliminate theory from their epistemology. Lewis and Kuhn argued, correctly, that this is impossible. The idea that theory was necessary for knowledge was not new to Lewis but is also present in the works of Henri Poincaré (1854 – 1912) who was duly reference by Lewis.

Another person Lewis influenced was Willard V. O. Quine (1908 – 2000), although Quine and Lewis did not agree. Quine is perhaps best known outside the realm of pure philosophy for the Duhem[2]-Quine thesis, namely that it is impossible to test a scientific hypothesis in isolation because an empirical test of the hypothesis requires one or more background assumptions. This was the death knell of any naïve interpretation of Sir Karl Popper's (1902 –1994) idea that science is based on falsification. But Quine's main opponents were the logical positivists. Popper was just collateral damage. Quine published a landmark paper in 1951: TWO DOGMAS OF EMPIRICISM. I would regard this paper as the high point in the discussion of the scientific method by a philosopher and it is reasonably readable (unlike Lew-

---

[1] I prefer Lewis.
[2] After Pierre Duhem (1861 – 1916).

is's The Mind and the World Order). Beside the Duhem-Quine thesis, the other radical idea is that observation underdetermines scientific models and that simplicity and conservatism are necessary to fill the gap. This idea also goes back to Poincaré and his idea of conventionalism—much of what is regarded as fact is only convention; conventions being necessary because observation underdetermines the models.

To a large extent my ideas match well with the ideas in Two Dogmas of Empiricism. Quine summarizes it nicely in that publication as:

> *The totality of our so-called knowledge or beliefs, from the most casual matters of geography and history to the profoundest laws of atomic physics or even of pure mathematics and logic, is a man-made fabric which impinges on experience only along the edges.*

and

> *The edge of the system must be kept squared with experience; the rest, with all its elaborate myths or fictions, has as its objective the simplicity of laws.*

Amen.

Despite Quine's fame in philosophical circles, the ideas in The Two Dogmas of Empiricism seem to be underappreciated. In a recent discussion of simplicity in science I came across, there was neither a single mention of Quine's work nor his correct identification of the role of simplicity—to relieve the under determination of models by observation. Sad.

Where philosophers have dropped the ball it was picked up by people in, of all places, management. Two people influenced by Lewis were Walter A. Shewhart (1891 – 1967) and Edwards Deming. It is said that Shewhart read Lewis's book fourteen times and Deming read it nine times. Considering how difficult that book is, it probably required that many readings just to comprehend it. Shewhart is regarded as the father of statistical process control, a key aspect of quality control. He also invented the control chart, a key component of statistical process control. Shewhart's 1939 book Statistical Method from the viewpoint of Quality Control[1] is a classic in the field but a large part of it was devoted to showing how his ideas are consistent with Lewis's epistemology.

In this 1939 book, Shewhart introduced the Shewhart Cycle, which was the first place where production is seen as part of a cycle—specifications, production, inspection, repeat—and not just a linear process that stops after inspection. Shewhart explicitly modeled his cycle after the hypothesis, experiment, and test cycle of the scientific method. The Shewhart cycle was modified by Deming and is sometimes called the Deming Wheel. Under its current name the Plan-Do-Check-Act (PDCA) cycle forms the basis of ISO management standards[2]. It is striking that when we want practical knowledge, even in business management, we turn to the scientific method.

Deming added the largely superfluous *act* step between inspection and specifications. It can easily be incorporated in the specification or plan stage (as it is in Shewhart's original cycle). But Deming was influenced by Lewis who regarded knowledge without the possibility of acting as impossible, hence the *act* step. This idea has become ingrained in ISO management standards under the rubric *continual improvement* (Clause 10 in the ISO standards).

Deming's 1993[3] book THE NEW ECONOMICS illustrated the extent to which he was influenced by Lewis. There he summarizes his approach to business management in his system of profound knowledge. This has four parts: knowledge of system, knowledge of variation, theory of knowledge and knowledge of physiology. The one that seems out of place is the third; why include theory of knowledge? Deming believed that this was necessary for running a company and he explicitly refers to Lewis's 1929 book. Making the reading of Lewis's book mandatory for business managers would certainly have the desirable effect of cutting down the number of managers. To be

---

[1] Deming edited and wrote an introduction to the book.
[2] See for example ISO Annex SL, Appendix 3.
[3] Indeed, written when he was 93.

fair to Deming, he does suggest starting in about the middle of the book.

As we have seen, there are two related unbroken chains: 1) Peirce, Lewis, Shewhart, Deming, ISO management standards and 2) Pierce, Lewis, Quine, my philosophical musings[1]. It reminds one of James Burke's TV program CONNECTIONS. Sir Karl Popper (1902 – 1994) with his idea of falsification may be the person many scientists use to justify how they work but Quine would probably be better and Quine's teacher, C.I. Lewis, through Deming, has provided the philosophic foundation for business management. Within the context of Definition 3) for knowledge both science and business have been very successful and were needed for you to access this essay. So, philosophy is not always useless, some of it is pragmatic.

2. ON THE NATURE OF SCIENCE

There are two very peculiar things about the scientific method: first, how late in the development of civilization it became mainstream, and second, that is there is no generally accepted definition of what it actually is, certainly not within the philosophical community.

Hints of the scientific method date back to the astronomy of ancient Babylon (c. 1000 BCE), to the philosophy of Thales (624 BCE – 546 BCE) of Miletus in ancient Greece, and to the experimentation of Frederick II (1194 – 1250) and Roger Bacon (c. 1214 – 1294) in medieval Europe. But it was only when Galileo (1564 – 1642) turned his telescope on the heavens in 1609 that it *took*. It was only then that the scientific method was finally on the road to becoming a dominant part of everyday culture. When Kepler (1571 – 1639), and especially Newton (1643 – 1727), consolidated Galileo's work, there was no turning back. As they say, the rest is history.

There have been various ideas put forth in the past for what science is: induction, verification, falsification, and various other 'tions'.

---

[1] Excuse me for putting my musings in the same chain as the works of these giants.

7

There have also been monstrosities like methodological naturalism; dogma masquerading as method. But all these have their critics and justly so. In the end, the current consensus in the philosophical community—to the extent there is a consensus—is that the scientific method, as a unified concept, does not exist. Strange as it may seem, there is this general idea that there is no such thing as the scientific method but that different fields of science use different unrelated methods.

The problem is that the scientific method is not what people, especially the philosophical community, expected. The philosophical community has concentrated on things like knowledge, explanations, truth, facts, naturalism, realism, and other such abstruse metaphysical concepts. Yet, they have missed the obvious—that science is something simpler, much simpler, namely model building; hence the drawing of the model boat among the equations on the cover of this book. The basic idea of the scientific method is summarized as[1]:

1. Begin with a set of observations.
2. Create a model to explain the observations. A model is not reality but a portrayal of it.
3. Use simplicity to select one preferred model from the many possible models that describe any set of observations.
4. Make testable predictions using the model.
5. Compare the predictions to observations not involved in constructing the model.
6. Use the comparison to assess, modify or reject the model. If the model is rejected create a new one.
7. Return to step three and repeat the cycle to achieve continual improvement.

This model-building view of science allows us to understand the scientific method in a simple, unified manner valid across the whole spectrum of scientific endeavors and to see the shortcomings of other views of science. This model-building approach also allows us to minimize the metaphysics required. Unfortunately, metaphysics can never be completely eliminated.

---

[1] The basic idea is from Roger Bacon.

Model building is not enough to specify the scientific method. You need two additional concepts: observations and parsimony. The models of science are constrained by observations, and judged by their ability to make correct predictions about future observations. Like a model boat, scientific models cannot be proven right or wrong—what sense does it make to claim a model boat is right? But we can certainly say which of two model boats is a more accurate representation of the original. Similarly with scientific models: we can say which of two models is more accurate at making correct predictions for observations. We do not have induction, verification, or falsification, but rather comparison. As Sir Karl Popper (1902 – 1994) pointed out, we have replaced certainty with continual improvement: the models of science are becoming more accurate over time.

Now, observations by themselves are not able to uniquely determine a model. An infinite set of models make the same set of predictions, the same way an infinite number of mathematical curves may be drawn through any finite set of points. But, once it is accepted that science is about model building and making predictions for observables, it becomes clear that adding frills—that don't change the predictions—is counterproductive. Thus we use parsimony or simplicity to make our observationally constrained models unique. It is the combination of simplicity and observations that fully constrain scientific models. Willard Quine (1908 – 2000) added conservatism but that is probably unnecessary.

Models do more than allow one to make predictions; they provide structure and meaning to the observations. This is the point missed by the logical positivists who wanted to go straight from the observations to the meaning. Thomas Kuhn (1922 – 1996), picking up ideas from C.I. Lewis (1883 – 1964), pointed out the folly of this with his idea of paradigms: the structures needed to give order to any field of endeavor. Thus we have the essence of the scientific method: observational constrained model building, with the meaning in the model.

3. STERILE NEUTRINOS AND THE PROBLEM OF EVIL

The thesis of this essay is not that sterile neutrinos are evil; though they may well be. Nor is it that physicists are evil though their significant others may, at times, think they are. Rather it is that falsification, a concept which strongly influences the physics community's self-image, is much more subtle than usually acknowledged. Karl Popper (1902 – 1994) proposed in his work on the philosophy of science that what distinguishes science is that its hypotheses or models can be falsified. This, he identifies as the demarcation criteria for determining what science is. One hears the question: *Is that falsifiable?* in laboratory-hallway conversations. However, the philosophy community has largely dismissed Popper's ideas. The main criticism is known as the Duhem-Quine thesis[1]: it is impossible to test a scientific hypothesis in isolation, because an empirical test of the hypothesis requires one or more background assumptions. Let's see how this works or doesn't work by considering two examples: one from classical theology and one from modern physics.

Below, we have a statement of the problem of evil (from WIKIPEDIA):

> *Is God willing to prevent evil, but not able?*
> *Then he is not omnipotent.*
> *Is he able, but not willing?*
> *Then he is malevolent.*
> *Is he both able and willing?*
> *Then whence cometh evil?*
> *Is he neither able nor willing?*
> *Then why call him God?*

The statement of the problem is first attributed to Epicurus (341 BCE – 270 BCE), predating Christianity. By the Popper demarcation criteria this would be a scientific hypothesis. If you accept the assumption that there is an omnipotent, omni-benevolent God, then there is the

---

[1] Named for Willard V. O. Quine (1908 –2000) and Pierre M. M. Duhem (1891 – 1916)

falsifiable prediction that evil does not exist. At first blush, it looks like the assumption has been falsified and God cannot be both omnipotent and omni-benevolent. But now the Duhem-Quine thesis kicks in. One of the many refutations of the problem of evil comes from Gottfried Leibniz (1646 – 1716). That's right—the opponent of Newton on the priority of the discovery of calculus. There was no separation of science and religion in those days.

Leibniz's argument was this is the best of all possible worlds: a position known as optimism and ridiculed in Voltaire's CANDIDE (published 1759). Putting that ridicule aside, let's look at the argument. Leibniz said that humans have only seen a very small part of the entirety of existence. In his theology, souls exist forever and so it would be premature to judge if this is the best possible world since we have seen such a small fraction of it. He has attacked the *whence cometh evil?* question, one of the background assumptions and not the main assumptions of omnipotence and omni-benevolence. According to Leibniz, the reason we do not understand where evil comes from is due to our limited knowledge and understanding. Once we have a better understanding, presumably in the afterlife, we will see where evil, or the appearance of evil, comes from. This approach used the Duhem-Quine thesis to get around the falsification. For the sake of argument let us accept his argument as valid. It avoids falsification, but at a price. While the initial statement meets Popper's demarcation requirement for science, Leibniz has removed any possibility of falsification. Any outcome is consistent with our limited understanding and the problem of evil has been removed from science to the realm of theology or philosophy. This is perhaps where it should be.

Now before physicists get too smug and think this is only a problem for theology, let us consider another problem: sterile neutrinos and neutrino oscillations. Here again we will see that falsification can be avoided but only at the price of moving from science to metaphysics

If we take the three known generations of particles, we have three neutrinos. Various experiments, at Super Kamiokande, Sudbury Solar Neutrino Observatory (SNO), and elsewhere, have shown that the neutrinos oscillate. Most of these experiments are consistent with three neutrinos oscillating among themselves. But there is a striking exception—the Liquid Scintillator Neutrino Detector (LSND) experiment at the Los Alamos National Laboratory. There are two principle ways to explain the LSND results, 1) the experiment is wrong, or 2) there is another type of neutrino, a so-called sterile neutrino.

Both of these approaches avoid having LSND falsify the main idea that what is being seen in the previous experiments is due to neutrino oscillations. Both reduce the ability to falsify predictions. If we simply assume that all experiments that disagree with the prevailing wisdom are wrong, we have nothing left and falsification becomes impossible. So we need a good reason to throw out experimental results we do not like. Way 2) also makes falsification more difficult—not impossible—but more difficult. Adding an extra neutrino means we need more parameters which can then be adjusted, meaning we have more wiggle room to explain inconsistent results.

To help decide between these two ideas a new experiment was carried out: Mini Booster Neutrino Experiment (MiniBooNE). The results of this experiment were inconsistent with the LSND results—*If* you assume only one sterile neutrino. But Duhem-Quine strikes again. Why not assume three sterile neutrinos, one for each generation of particles? We avoid falsifying the hypothesis that the LSND is due to sterile neutrinos, but at the price of introducing more parameters and making the new model less falsifiable. If we add additional sterile neutrinos at will, we can completely avoid any possibility of falsification. In doing so we fail Popper's demarcation criteria and the model becomes unfalsifiable. In general, as we add more parameters we move down the slippery slope to unfalsifiable, from science to metaphysics. (I avoid saying from science to theology.) But, what happens when, in avoiding falsification, we make the model more falsifiable? Well that's another paradigm.

4. PARADIGMS

Thomas Kuhn (1922 – 1996) began his career as a physicist but then, as a post-doctoral fellow, went over to the dark side and became a philosopher. It is for his work on the dark side that he became famous. Normally one assumes that when a scientist starts doing philosophy it is a sign of senility, but in his case it was too early in his career and his insights were actually useful (Yes, philosophy can occasionally be useful). His main contribution was his introduction of the idea of the paradigm[1]. A paradigm is the set of interlocking assumptions and methodologies that define a field of study. It provides the foundation for all work in the field and a common language for discourse. It is the fundamental model for the field and in historical studies is sometimes referred to as the controlling narrative.

The frequent use of the phrase, *paradigm change,* makes you think that all paradigms do is change. But the idea of the paradigm is actually subversive—it helped undermine the *received view* of what science is and still undermines experimentalists' attempts to eliminate theory (Which can't be done, by the way!). Full disclosure: I am, or rather WAS, a theorist. Administration, which is my current lot, is even farther to the dark side than philosophy.

The concept of paradigm was introduced in contradistinction to the ideas of positivism that defined the *received view.* The positivists tried to work directly with observations and eliminate all metaphysics or model dependence. Kuhn, on the other hand, claimed the observations themselves are theory laden or model dependent. You cannot, as a matter of principle, eliminate the metaphysics because the observations, or at least their interpretations, depend on the theory, model, or paradigm. The paradigm sets the framework that gives meaning to the observations and frames the very questions that are considered worthy of addressing. Examples of paradigms would be Aristotelian physics, classical physics, the standard model of particle physics, or the modern synthesis of evolution.

While paradigms do *more* than change, they do indeed change and when they do all—oops I cannot say that!—all *heck* breaks loose. Things one thought one knew and could rely on suddenly go poof. This going *poof* was what the positivists tried and failed to get around by eliminating the models and working directly with the observations.

---

[1] The idea was already implicit in the work of C.I. Lewis (1883 – 1964), but Kuhn made it explicit and popularized it.

As Albert Einstein (1879 –1955) pointed out, when paradigms change, it tends to be the most central parts of the previous paradigm that are eliminated. In Aristotelian physics, it was the fixed earth and the perfect heavens that Galileo destroyed with his telescope.

Classical mechanics is built on Euclidean three-dimensional space and well-defined trajectories. Special and general relativity eliminated Euclidean geometry, and string theory, if correct, means space is not three-dimensional. Quantum mechanics eliminated the well-defined trajectories. This still causes some people sleepless nights but does not bother me since most of the time I do not know where I am or let alone where I am going. Evolution wreaked havoc with the concept of fixed species. Before continental drift was accepted, a central concept of geology was the continents did not move. The examples are endless.

A side effect of this is that one cannot depend on the contents of the present theories or models to have any direct connection with reality. The ether (electromagnetism), caloric (heat), phlogiston (fire), and mal air (medicine) that at one time were essential parts of the understanding of how the universe works were eliminated by new improved models. There is no guarantee that the contents of the current models will not be similarly eliminated. Maybe we will find quarks disappearing or more likely, the idea of time itself disappearing since time is apparently more fundamental.

So what is science and what is it good for if the basic concepts keep changing? Well now, that is a good question.

5. FLYING REINDEER AND THE NATURE OF SCIENCE

In Honour of Bob Moore (1935-2011) Professor Emeritus, McGill University

A few years ago I published an article on how science works in Physics in Canada. Since I failed philosophy in university, I refuse to call it philosophy of science. The response was like the advertising slogan for Keith's India Pale Ale (a Nova Scotian brew): *Those that like it, like it a lot.* One of the letters I received on the paper came from Bob Moore, a colleague from my McGill days. He indeed liked it a lot. However he did point out one aspect of the how science works that I had overlooked. The paper was already getting rather lengthy, and the usual comment was not that I left things out but had put too much in.

Now, Bob was a Newfoundlander or Newfie for short. Like many Newfies he had the gift to tell a good story. Thus I considered it the ultimate compliment when he said on reading parts of my paper, *I could have written that.* What I had left out, he noted, was the idea that scientists are more like gamblers or bookies than priests (the holders of eternal truths) or Gnostics (processors of secret knowledge). Bob sent me an essay he had written on the topic which I will now mangle and summarize.

Consider the thesis that reindeer can fly. How do we test that idea? Well, take some reindeer to the top of a tall building and push them off. There goes Dasher; Splat, Dancer: Splat, … Blitzen: Splat. Okay, have we falsified the thesis that reindeer can fly? No we have only shown that those reindeer in that particular instance did not fly. Perhaps they could have but chose not to. Perhaps they could not but there are others that could. Perhaps it was the wrong time of year. We have proven nothing, but only given another example of the Duhem-Quine thesis that any potential falsification can be gotten around. Bob's point was that we were really asking the wrong question. The correct question is[1]: How would you bet on Rudolph?

In this uncertain world, the role of science is much like that of bookies—to set the odds of what will happen, not to discover eternal truths. Will the sun rise tomorrow? Highly probable. Will the LHC find supersymmetry? Unlikely. Will Vancouver ever win the Stanley Cup? Unlikely. (Toronto? Forget it.) Can reindeer fly? Very unlikely. In a very real sense all we ever do in science to determine probability.

---

[1] The change in the question is the essence of the switch to the pragmatic view of knowledge. While the reindeer example is originally from James Randi (b. 1928), the question is pure Bob Moore.

So why is science so successful? Because, as any professional gambler knows, playing by the odds gets you more wins than not playing by the odds. Playing by hunches and hoping you luck out will *work* occasionally, but not in the long run. And if you play against the laws of physics, the odds against lucking out can be very long. The odds that the atomic fine structure constant is between 0.00729735242 and 0.00729735271 are estimated to be about a billion to one. And the odds for the conservation of momentum in physics are so high as to be incalculable—but not infinite. Some physicists think that the law could break down in the extreme gravitational field of a neutron star orbiting a black hole.

For scientists who have to deal with things as complex as the health and wellbeing of human beings, the odds can be hard to determine. Yet for our health and wellbeing it is often important to try to do so. Not doing so leads to false beliefs and superstitions, like using apricot pits to cure cancer. The most ludicrous I have heard of is a hockey player who before each game dipped his stick in a toilet bowl. Perhaps the Vancouver Canucks should have tried that. Ah, perhaps they did!

Beliefs, such as the power of laetrile (usually made from apricot pits) `to cure cancer and the aphrodisiacal power of ground rhinoceros horn can be detrimental to our health and wellbeing, and certainly to the rhinoceroses'. So when someone tells you about a study that shows evidence of flying saucers, or of mental telepathy, or that apricot pits can cure cancer, practice a little *I come from Missouri* and ask *How much do you want to bet?*

However, in addition to death and taxes, one other thing is certain: We will miss Bob's humour, newfie stories, and insights. Good-bye old friend, so long.

6. DON'T TAKE SPORTING RESULTS SERIOUSLY

A colleague of mine is an avid fan of the New York Yankees baseball team. At a meeting a few years ago, when the Yankees had finished first in the American league regular season, I pointed out to him that the result was not statistically significant. He did not take kindly to the suggestion. He actually got rather angry! A person, who in his professional life would scorn anyone for publishing a one sigma effect, was crowing about a one sigma effect for his favorite sports team. But then most people do ignore the effect of statistical fluctuations in sports.

In sports, there is a random effect in who wins or loses. The best team does not always win. In baseball where two teams will frequently play each other four games in a row over three or four days, it is relatively uncommon for one team to win all four games. Similarly a team at the top of the standings does not always beat a team lower down. As they say in sports: on any given day, anything can happen. Indeed it can and frequently does.[1]

Let us return to American baseball. Each team plays 162 games during the regular season. If the results were purely statistical with each team having a 50% chance of winning any given game, then we would expect a normal distribution of the results with a spread of sigma=6.3 games. The actual spread or standard deviation for the last few seasons is closer to 11 games. Since spreads are added in quadrature most of the result is due to real differences. However, moving from the collective spread to the performance of individual teams, if a team wins the regular season by six games or one sigma, as with the Yankees above, there is a one in three chance that it is purely a statistical fluke. For a two-sigma effect, a team would have to win by twelve games or for a three-sigma effect by eighteen games. The last would give a confidence level of over 99% that the winner won based on skill, not due to a statistical fluctuation. When was the last time any team won by eighteen games?

For particle physics we require an even higher standard—a five sigma effect to claim a discovery. Thus a team would have to lead by 30 games to meet this criterion. Now my colleague from the first paragraph suggested that by including more seasons the results become more significant. He was right of course. If the Yankees finished ahead by six games for thirty-four seasons in a row that would be

---

[1] With the expected frequency of course.

five-sigma effect. From this we can also see why sports results are never published in Physical Review with its five-sigma threshold for a discovery—there has yet to be such a discovery. To make things worse for New York Yankees' fans they have already lost their chance for an undefeated season this year.

In other sports the statistics are even worse. In the National Hockey League (NHL), teams play eighty-two games and the spread in win-loss expected from pure chance is sigma=4.5. The actual distribution for 2011 was sigma=6.3. The signal due the variation in the individual team's ability is all in the 1.8 difference in the sigmas. Perhaps there is more parity in the NHL than in Major League Baseball. Or perhaps there is not enough statistics to tell. Speaking of not telling, in 2011 the Vancouver Canucks finished with the best record for the regular season, two games ahead of the New York Rangers and three games ahead of the St. Louis Blues. Only a fool or a Vancouver Canucks fan would think this ordering was significant and not just a statistical fluctuation. In the National Football League last year, 14 of the 32 teams were within two sigma of the top. Again much of the spread was statistical. It was purely a statistical fluke that the Seattle Seahawks did not win the Super Bowl as they should have.

Playoffs are even worse (this is why the Canucks have never won a Stanley Cup). Consider a best of seven game series. Even if the two teams are equal, we would expect that the series would only go four games one in every eight (two cubed[1]) series. When a series goes the full seven games one might as well flip a coin. Rare events, like one team winning the first three games and losing the last four, are expect to happen once in every sixty-four series and considering the number of series being played it is not surprising we see them occasionally.

Probably the worst example of playoff madness is the American college basketball tournament called, appropriately enough, March Madness. Starting with 64 teams or 68 depending on how you count, the playoffs proceed through a single elimination tournament. With over 70 games it is not surprising that strange things happen. One of

---

[1] Not two to the fourth power because one of the two teams has to win the first game and that team has to win the next three games.

the strangest would be that the best team wins. To win the title the best team would have to win six straight games. If the best team has on average a 70% chance of winning each game they would only have a 12% chance of winning the tournament. Perhaps it would be better if they just voted on who is best.

But you say they would never decide a national championship based on a vote. Consider American college football. Now that is a multi-million dollar enterprise! Nobel Laureates do not get paid as much as US college football coaches. They do not generate as much money either. So what is more important to American universities—sports or science?

In the past, the US college national football champions were decided by a vote of some combination of sports writers, coaches and computers. Now that combination only decides who will play in the championship game or a four-team playoff. The national champion is ultimately decided by who wins that one final game. Is that better than the old system? More exciting, but as they say: on any given day anything can happen. Besides, sports is more about deciding winners and losers rather than who is best.

7. TRUTH VERSUS CONVENIENT HYPOTHESIS

July 17, 2012 marked the 100th anniversary of the death of Henri Poincaré (1854 – 1912). Henri who? You may well ask. He was just an ordinary, run of the mill genius: engineer, mathematician, physicist and philosopher. Believe me, you have to be a genius to be both an engineer and a philosopher. As an engineer he carried out the investigation into the 1879 Magny mining disaster. As a physicist he introduced the ideas of relativity, using light beams to synchronize watches and is attributed by some as the real inventor of special relativity. In mathematics, he is known for the Poincaré conjecture in topology and the beginning of chaos theory. But it is his philosophy that I want to explore here.

Poincaré was the father of what is known in philosophical circles as conventionalism: what others regarded as laws derived by the scientific method he regarded as convenient hypotheses or conventions. Two examples he gave are the geometry of space-time and Newton's laws of motion. The following is typical of this thought (Science and Hypothesis, original version 1902, English version 1905):

> *Whether the ether exists or not matters little—let us leave that to the metaphysicians; what is essential for us is, that everything happens as if it existed, and that this hypothesis is found to be suitable for the explanation of phenomena. After all, have we any other reason for believing in the existence of material objects? That, too, is only a convenient hypothesis; only, it will never cease to be so, while some day, no doubt, the ether will be thrown aside as useless.*

For him, even the existence of material objects[1] was just a convenient hypothesis. This might seem extreme but he was spot on about the ether. Although the ether was a cornerstone of physics in 1904, it was demolished by Einstein's paper the following year.

To put more flesh on the idea of conventionalism and show that it is not ridiculous, consider the notion of a fixed earth: this is the old Copernicus vs Ptolemy or Galileo vs the Catholic Church story. Poincaré would have regarded it as a convention that the earth is not stationary. Ah, you say, did not Galileo give the definitive answer with his telescope? Not at all. While the telescope did get rid of the Ptolemaic system, Tycho Brahe (1546 – 1601) came up with an alternative: a fixed earth orbited by the sun, and the other planets then orbiting the sun (the Duhem-Quine thesis strikes again). So, is there any way to settle the question of the absolute motion of the earth? No, all we can measure is the motion of earth with respect to something else: the sun, the center of the galaxy, the fixed stars, the cosmological three degree microwave background or some other specified inertial frame. Take any arbitrary frame you like (I prefer the one where I am the center of the universe, your mileage may vary) and there is a well-defined mathematical transformation from any of the previously mentioned

---

[1] Willard V. O. Quine (1908 – 200) presented similar ideas about 50 years later, calling them myths or fictions.

frames to this new one. We can use that transformation to write all the laws of physics consistently in the new frame.

This is not relativity—either special or general—which has the same laws in different frames. Here the laws are different in the different frames—different but well defined. The fixed earth description has some peculiarities:

- The laws of physics depend on the distance from the center of the earth. But hey, since we have assumed the earth is the center of the universe this is good and proper.
- There are forces, generated by the above-mentioned mathematical transformation, that depend on the behavior of cosmological microwave background (perhaps Mach's principle).

The choice of frame is due to convenience. Tycho Brahe's model lost to Copernicus's due to the difficulty of using it for planetary motion. However for the motion of stars within the galaxy, we do not use a sun-centered frame but a galaxy-centered frame and for cosmological calculations we use a still different frame. But the most widely used frame of all is that of a fixed earth. Ptolemy rules! Good models never die. Consider giving directions from TRIUMF to the UBC physics department (about a mile away, to use obsolete units). Now in a sun-centered frame, UBC is rapidly moving due to the rotation of the earth and its orbital motion. Thus the directions would have to be time-dependent and even a small mistake in following them would have the person lost in space. Even worse than not taking that left *to-in* in *Albakoikie* (Bugs Bunny, HERR MEETS HARE, 1945).

That's classical mechanics, in quantum mechanics things are even worse. What is well defined in classical mechanics may become ambiguous in quantum mechanics. A prime example is the unitary transformation that changes all kinds of things around but leaves observations untouched. One of the most commonly used concepts in both classical and quantum mechanics is the potential. The gravitational potential holds us on the earth and the earth in its orbit (sun-centered frame). The nuclear potential holds the nucleus together, but, but, but, the potential can change dramatically under unitary transformations, from strongly repulsive at short distances to highly non-local. The

21

actual form used is no more and no less than a convention—a convenient hypothesis.

How do you choose the best convention? That is simplicity itself (see the next essay).

8. SCIENCE AND SIMPLICITY

One of the leading naturalists of the 19th century was Phillip Henry Gosse (1810 – 1888). He spent time in Newfoundland and Ontario where he cataloged insect species, among other things. On his return to England he published CANADIAN NATURALIST (1840) and became one of the leading popularizers of natural science of his day. He invented, or at least made practical, the marine aquarium with his 1854 book, THE AQUARIUM, which initiated a craze for aquariums. He was elected a Fellow of the Royal Society in 1856. He even communicated with Charles Darwin (1809 – 1882) over the study of orchids. Altogether, a first rate naturalist.

He was also an extremely devout Christian, a member of the Plymouth Brethren. It bothered him that the geological record implied an age of the earth much older than the age given in the Bible. Gosse's solution was given in a book called OMPHALOS: AN ATTEMPT TO UNTIE THE GEOLOGICAL KNOT (published in 1857, two years before Darwin's famous book). The title, Omphalos, is the Greek word for belly button; as in, did Adam have a belly button? The book itself is rather tedious with page after page of examples, but the first part is well worth a read.

Okay, anyone who has read my previous essays knows what is coming next. That's right, the Duhem-Quine Thesis: the idea that one can always avoid falsification by adjusting an auxiliary hypothesis. In this case, it's a doozie. Gosse suggested God created the universe only six thousand years ago but in such a manner that it is indistinguishable from an old one: light created in transit from stars, fossils created in rocks, etc. Gosse made two distinct points:

- Any act of special reaction implies a false history. A created chicken implies an egg that did not exist or vice versa. A cre-

22

ated tree would have rings not due to growth. Dinosaur fossils, however, do seem a bit extreme.

- The universe could be any age and it is only an outside reference (e.g. the Bible) that allows one to determine the actual age.

Needless to say the book was not a hit, denounced by Christians (God is a deceiver? Bah humbug) and ignored by scientists (Am I studying an illusion? Bah humbug). Even a name change did not save it, and in 1869 the remaining copies were sold for scrap. However, it has significant epistemological implications that has led to some modern parodies like last Thursdayism; the idea that the earth was created last Thursday but only appears older.

Okay, one may be able to argue about Omphalosism, but Last Thursdayism is clearly absurd. But why? What postulate of the scientific method does it violate? It, by construction, makes all the same predictions as the standard models. Thus, it cannot be eliminated by the appeal to observation, the touchstone of the scientific method. I would suggest the only criterion is simplicity. It is simplicity that eliminates Omphalosism. Extra complexity has been added to the model with no gain in the ability to make predictions.

Simplicity is like air; it is so ubiquitous that one tends to forget it is there. But it is there; from William of Occam (1238 – 1348) (Occam's razor: Entities should not be multiplied unnecessarily), to Isaac Newton (We are to admit no more causes of natural things than such as are both true and sufficient to explain their appearances) to Steven Weinberg (You may use any degrees of freedom you like to describe a physical system, but if you use the wrong ones, you'll be sorry![1]).

In my last essay, I introduced the idea of conventionalism—that what is frequently regarded as truth is but a convention. Weinberg's degrees of freedom are an example. But how is the convention chosen? I would suggest simplicity plays the dominate role. We use an earth-based frame or a sun-based frame due to simplify or ease of use. We

---

[1] From: ASYMPTOTIC REALMS OF PHYSICS (ed. by Guth, Huang, Jaffe, MIT Press, 1983)

23

use a nuclear potential that is weak at short distances like Vlow k rather than the traditional potentials with strong short-range repulsion due to simplicity and ease of use. Only this, and nothing more (to quote Edgar Allan Poe's (1809 – 1849) poem THE RAVEN).

These considerations of the Omphalos thesis and Last Thursdayism (some heretics believe in Last Tuesdayism, but we will excommunicate them) make the emphatic point that simplicity is necessary, absolutely necessary. Otherwise one can multiply hypotheses without limit and get bogged down in futile arguments (Last Thursdayism vs Last Tuesdayism, Vlow k vs the traditional potentials). We draw smooth curves through data points rather than wiggly ones, again due to simplicity. Simplicity all the time and everywhere. Simplicity rules!

While simplicity is crucial, in the end it leads us astray. Newtonian dynamics was replaced by more complex models (relativity and quantum mechanics), fixed continents were replaced by the more complex idea of continental drift, animals reproducing after their kind was replaced by the complexities of evolution, and on it goes with simple paradigms being replaced by more complex ones. Mr. Kuhn[1] meet Mr. Murphy[2], Mr. Murphy meet Mr. Kuhn.

9. WHAT IS SCIENCE?

What is science? How about: Model-dependent realism? I'm aware that this succinct definition sounds like an oxymoron: if it is model dependent how can it be realism? Model-dependent realism comes from the book THE GRAND DESIGN (2010) by Stephen Hawking (b. 1942) and Lenard Mlodinow (b. 1954). The essential idea is that science is model building, but the internal aspects of the models are not significant; reality is model dependent. All our theories, laws, hypotheses are models, models for how the universe works. Nothing more and nothing less. What are the criteria for choosing a good model? There is really only one criterion: agreement with observation.

---

[1] Thomas Kuhn introduced the idea of paradigm change.
[2] Everyone has run afoul of Murphy's Law.

Yet, as discussed in a previous essay there is also simplicity. But this is really a corollary of the first point. If the only criterion is agreement with observation, there is nothing to be gained by encumbering your model with frills that are unnecessary for making predictions for observations. Get rid of them.

Model-dependent realism is not all that different from the conceptual pragmatism of C.I. Lewis (1883 – 1964). There what is real is determined by the pragmatically determined concept (model) of the real. Another very similar idea for how knowledge is acquired goes by the name Critical Realism. There seems to be a theorem that all approaches to science should have realism in the name, regardless of how appropriate it is. Personally, I would just call it model building and leave the realism out. But who am I to argue with Steven Hawking (I will anyway. And the ultimate authority, WIKIPEDIA, does say it is controversial).

Critical realism traces its roots back to a chemist, Michael Polanyi (1891 – 1976), but I first read about it in the book THE NEW TESTAMENT AND THE PEOPLE OF GOD (1992) by N.T. Wright (then the Anglican Bishop of Durham). The description given there is devoted mainly to history. It is similar to model-dependent realism but puts more credence on the internals of the models. Even though Polanyi takes precedence for this theory, Wright's exposition is one of the best descriptions of how science works that I have read. It will seem strange to many people that in a book that attacks the idea of God (THE GRAND DESIGN) and a conservative Christian theologian propose similar models for obtaining knowledge. But it really isn't that strange. There is a common foe: postmodernism.

Postmodernism is the idea that all points of view are equally valid. This is anathema to both scientists and conservative Christians. Both scientists and Christians want to propose a criteria that separates the sheep from the goats, the wheat from the chaff, or the valid models from the dung heap of postmodernism (the reader may gather that I am not a great fan of postmodernism either).

How is the separation of sheep from goats done? Contrary to the positivists' claim, models cannot be verified and contrary to Popper, they can be not be falsified either (the dreaded Duhem-Quine thesis) but they can be compared. Established models (the word I prefer to theory, law, or controlling narrative) are rarely abandoned because they simply disagree with observation. Rather they are abandoned because another model does a better job. To paraphrase the American gun lobby: Observations do not kill models, models do. The ether was only abandoned after Einstein proposed the special theory of relativity; the Michelson-Morley experiment was not sufficient. Michelson and Morley may have provided the ammunition, but it was Einstein who pulled the trigger. His defense lawyer would argue that the ether was not shot, but rather had its throat slit with Occam's razor. Einstein did not prove that the ether did not exist. Rather he showed that the ether hypothesis, like the Omphalos hypothesis, has no predictive power, and in the end, it was eliminated by appeals to simplicity. This would not have surprised Henri Poincaré (1854 – 1912).

You can go through example after example and see observation deciding between or among competing models. The big bang model of the universe beat out the steady state model because it predicted the three degree microwave background, the quark model beat out rivals with the discovery of the J/Psi particle, and continental drift beat out the fixed continent model when the seabed of Atlantic was explored in detail. In every case, it is model combat and the one that was best at predicting new phenomena won. If you want to become rich and famous (or at least win a Nobel Prize) come up with a model that makes a striking prediction (none of this postdiction nonsense) and have it confirmed by observation. That is the tricky part (bribing experimentalists does not count).

What do the internals of models mean? Are they really quite as meaningless as the model-dependent realism implies or do they have meaning, as the critical realism suggests. I take an intermediate approach and follow Poincaré. I define a thing to exist if the math goes through as if the thing did exist. What's good enough for Poincaré is good enough for me.

10. THEORETICAL PHYSICS IS A QUEST FOR SIMPLICITY

Theoretical physics, simplicity? Surely the two words do not go together. Theoretical physics has been the archetypal example of complicated since its invention. So what did Frank Wilczek (b. 1951) mean by that statement[1] quoted in the title of this essay? It is the scientist's trick of taking a well-defined word, such as *simplicity*, and giving it a technical meaning. In this case, the meaning is from algorithmic information theory. That theory defines complexity (Kolmogorov complexity[2]) as the minimum length of a computer program needed to reproduce a string of numbers. Simplicity, as used in the title, is the opposite of this complexity. Science, not just theoretical physics, is driven, in part, but only in part, by the quest for this simplicity.

How is that you might ask. This is best described by Gregory Chaitin (b. 1947), a founder of algorithmic information theory. To quote[3]:

*This idea of program-size complexity is also connected with the philosophy of the scientific method. You've heard of Occam's razor, of the idea that the simplest theory is best? Well, what's a theory? It's a computer program for predicting observations. And the idea that the simplest theory is best translates into saying that a **concise** computer program is the best theory. What if there is no concise theory, what if the most concise program or the best theory for reproducing a given set of experimental data is **the same size** as the data? Then the theory is no good, it's cooked up, and the data is incomprehensible, it's random. In that case the theory isn't doing a useful job. A theory is good to the extent that it compresses the data into a much smaller set of*

---

[1] From "This Explains Everything", Ed, John Brockman, Harper Perennial, New York, 2013
[2] Also known as descriptive complexity, Kolmogorov–Chaitin complexity, algorithmic entropy, or program-size complexity.
[3] From THINKING ABOUT GODEL AND TURING: ESSAYS ON COMPLEXITY, 1970-2007, page 119.

theoretical assumptions. The greater the compression, the better!—That's the idea...

In many ways this is quite nice; the best theory is the one that compresses the most empirical information into the shortest description or computer program. It provides an algorithmic method to decide which of two competing theories is best (but not an algorithm for generating the best theory). With this definition of best, a computer could do science: generate programs to describe data and check which is the shortest. It is not clear, with this definition, that Copernicus was better than Ptolemy. Their competing approaches to planetary motion had a similar number of parameters and accuracy.

There are many interesting aspects of this approach. Consider compressibility and quantum mechanics. The uncertainty principle and the probabilistic nature of quantum mechanics put limits on the extent to which empirical data can be compressed. This is the main difference between classical mechanics and quantum mechanics. Given the initial conditions and the laws of motion, classically the empirical data is compressible to just that input. In quantum mechanics, it is not. The time, when each individual atom in a collection of radioactive atoms decays, is unpredictable and the measured results are largely incompressible. Interpretations of quantum mechanics may make the theory deterministic, but they cannot make the empirical data more compressible.

Compressibility highlights a significant property of initial conditions. While the data describing the motion of the planets can be compressed using Newton's laws of motion and gravity, the initial conditions that started the planets on their orbits cannot be. This incompressibility tends to be a characteristic of initial conditions. Even the initial conditions of the universe, as reflected in the cosmic microwave background, have a large random non-compressible component—the cosmic variance. If it wasn't for quantum uncertainly, we could probably take the lack of compressibility as a definition of initial conditions. For the universe, the two are the same since the lack of compressibility in the initial conditions is due to quantum fluctuations but that is not always the case.

The algorithmic information approach makes Occam's razor, the idea that one should minimize assumptions, basic to science. If one considers that each character in a minimal computer program is a separate assumption, then the shortest program does indeed have the fewest assumptions. But you might object that some of the characters in a program can be predicted from other characters. However, if that is true, then the program can probably be made shorter. This is all a bit counterintuitive since one generally does not take such a fine-grained approach to what one considers an assumption.

The algorithmic information approach to science, however, does have a major shortcoming. This definition of the best theory leaves out the importance of predictions. A good model must not only compress known data, it must predict new results that are not predicted by competing models. Hence, as noted in the introduction, simplicity is only part of the story.

The idea of reducing science to just a collection of computer programs is rather frightening. Science is about more than computer programs[1]. It is, and should be, a human endeavour. As people, we want models of how the universe works that humans, not just computers, can comprehend and share with others. A collection of bits on a computer drive does not do this.

11. IS SCIENCE JUST ANOTHER RELIGION?

Modern science has assumed many of the roles traditionally played by religion and, as a result, is often mistaken for just another religion; one among many. But the situation is rather more complicated and many of the claims that science is not a religion come across as a claim that science is *The One True Religion*. In the past, people have used religion and religious concepts to supply answers to the basic questions of how the universe originated, how people were created, what determines morality, and how humans relate to the rest of the universe. Science is now slowly but surely replacing religion as the

---

[1] In this regard, I have a sinking feeling that I am fighting a rearguard action against the inevitable.

framework to answer these questions. Thus we have: the visible universe originated with the big bang, humans arose through evolution, morality arose through the evolution of a social ape and humans are a mostly irrelevant part of the larger universe. One may not agree with science's answers but they exist and influence even those who do not explicitly believe them.

More importantly, through answering questions like these, religion has formed the basis for people's worldview, their overall perspective from which they see and interpret the world. Religious beliefs and a person's worldview were frequently so entangled that they are often viewed as one and the same thing. In the past this was probably true, but in this modern day and age, science presents an alternative to religion as the basis for a person's worldview. Therefore science is frequently seen as a competing religion, not just the basis of a competing world view. Despite this, there is a distinct difference between science and religion and it has profound implications for how they function.

The prime distinction was recognized at least as far back as Thomas Aquinas (1225 – 1274). The idea is this: Science is based on public information while religion is based on private information, information that not even the NSA can spy on. Anyone can, if they wait long enough, observe an apple fall as Sir Isaac Newton (1642 – 1727) did, but no one can know by independent observation what Saint Paul (c. 5 – c. 67) saw in the third heaven. Anyone sufficiently proficient in mathematics can repeat Albert Einstein's (1879 – 1955) calculations but no one can independently check Joseph Smith's (1805 – 1844) revelations that are the foundation of Mormonism, although additional private inspiration may, or may not, support them. As a result of the public nature of the information on which science is founded, science tends to develop consensuses which only change when new information becomes available. In contrast, religion, being based on private information, tends to fragment when not constrained by the sword or at least the law. Just look at the number of Christian denominations and independent churches. While not as fragmented as Christianity, most major religions have had at least one schism. Even secularism, the none-of-the-above of religion, has its branches, one for example belonging to the new atheists.

The consensus-forcing nature of the scientific method and the public information on which it is based lead some to the conclusion that science is based on objective reality. But in thirty years of wandering around a physics laboratory, I have never had the privilege of meeting Mr. Objective Reality—very opinionated physicists, yes, but Mr. Objective Reality, no. Rather, science is based on two assumptions:

1. Meaningful knowledge can be extracted from observation. While this may seem self-evident, it has been derided by various philosophers from Socrates on down.
2. What was observed in the past can be used to predict what will happen in the future.

Science and religion are, thus, both based on assumptions but differ in the public versus private nature of the information that drives their development. This difference in their underlying epistemology means that their competing claims cannot be systematically resolved; they are different paradigms. Both can, separately or together, be used as a basis of a person's worldview and it is here that conflict arises. People react rather strongly when their worldview is challenged and the competing epistemologies both claim to be the only firm basis on which a worldview can be based.

12. MATHEMATICS: INVENTED OR DISCOVERED?

The empirical sciences, like physics and chemistry, are partially invented and partially discovered. Although the empirical observations are surely discovered, the models that describe them are invented through human ingenuity. But what about mathematics which is based on pure thought? Are its results invented or discovered?

Not surprisingly there are different views on this topic. Some people maintain that mathematical results are invented; others claim that they are discovered. Is there a universe of mathematical results just waiting to be discovered or are mathematical results invented by mathematicians only to disappear, like a fairy tale, when mathematicians vanish in the heat death of the universe? Invented or discovered? Perhaps some results are invented and others discovered. There is, however, a third view, namely that mathematics is a game played by manipulating symbols according to well-defined rules. At some level this is probably true. All those who prefer Monopoly®, put up your hands!

Bertrand Russell (1872 – 1970) and Alfred Whitehead (1861 – 1947) tried to derive mathematics from logic. The result was the book: PRINCIPIA MATHEMATICA (1910), a real tour de force. Their derivation still required axioms or assumptions beyond pure logic and it has been questioned on other grounds. An alternate to this approach is set theory, in particular based on the Zermelo-Fraenkel axioms, with the axiom of choice. And an alternate to that is category theory. Whatever all that is. It is certainly very technical. The quest for foundations of mathematics and even logic, like the quest for the Holy Grail, is probably never ending. But the question remains: were logic and set (category) theory, themselves, invented or discovered?

Let us look at things more simply. Historically, mathematics probably arose empirically: two stones plus two stones equals one stone plus three stones. Then it was realized that this holds for any tokens, stones, bushels of wheat or sheep. The generalization from specific examples to the generic 2+2=1+3 could be considered an early example of the scientific method: generalizing from specific examples to a general rule. But one plus one does not always equal two. Consider a litre of liquid plus a litre of liquid. If one is water and the other alcohol, the result is less than two litres if they are put in the same container. Adding one litre of water to one litre of concentrated sulfuric acid is even more interesting[1].

Multiplication is also easy to demonstrate with counters. Division is a bit more problematic but if we think of dividing a bushel of wheat into equal parts the idea of fractions is quite natural. Dividing a sheep is messier. Subtraction however leads to a problem: negative numbers. Naively, we cannot have fewer than zero stones but subtraction can lead to that idea. So were negative numbers invented or discovered? We can finesse the problem of negative numbers by saying that nega-

---

[1] Do not try this at home.

tive numbers correspond to what we owe. If I have minus three stones it means I owe someone three stones.

Thus thinking of stones and bushels of wheat, we can understand the rational numbers, numbers written as the ratio of two whole numbers. The Pythagoreans in ancient Greece would have claimed that is all there is. Then came the thorny problem of the square root of two? This arises in connection with the Pythagorean Theorem. Some poor sod showed that the square root of two could not be written as the ratio of two whole numbers and was thus irrational. He was thrown into the sea for his efforts[1]. The square root of two does not exist in the universe of numbers discovered using stones, sheep, and bushels of wheat. Is it possible to have square root of two stones? Was it invented to make the Pythagorean Theorem work or was it discovered?

The example of the square root of minus one is even more perplexing. We can think of the square root of two as an extra number inserted between 1.414 and 1.415. But there is no place to insert the square root of minus one. So again the question arises: was it invented or discovered? Perhaps it is best to say it was assumed. That is, assume the square root of minus one can be treated like a normal number[2] and see what happens: a lot of good things as it turned out, but does that mean it exists in any real sense. Perhaps it is just a useful fiction.

Nevertheless, mathematics has developed, discovering or inventing new results. As a phenomenologist, I would say we do not have enough information to assert if mathematics was invented or discovered. If we could contact extra-terrestrial mathematicians, it would be interesting to see if their mathematics was different or the same as ours. If it was different, that would be a strong indication that mathematics is invented. Or less black and white, the difference between terrestrial and extra-terrestrial mathematics would tell us the extent to which mathematics is discovered or invented.

---

[1] At least that is the legend.
[2] $\sqrt{(-1)} + \sqrt{(-1)} = 2\sqrt{(-1)}$, etc.

In any event, mathematics is a very interesting game, whether based on set theory or category theory, whether discovered or invented, and certainly more profitable than Monopoly®[1] in the long run.

13. KNOWLEDGE AND HIGGS BOSON

This essay makes a point that is only implicit in most of my other essays—namely that scientists are arro—oops that is for another post. The point here is that science is defined not by how it goes about acquiring knowledge but rather by how it defines knowledge. The underlying claim is that the definitions of knowledge as used, for example, in philosophy are not useful and that science has the one definition that has so far proven fruitful[2]. No, not arrogant at all.

The classical concept of knowledge was described by Plato (428/427 BCE – 348/347 BCE) as having to meet three criteria: it must be justified, true, and believed. That description does seem reasonable. After all, can something be considered knowledge if it is false? Similarly, would we consider a correct guess knowledge? Guess right three times in a row and you are considered an expert—but do you have knowledge? *Believed*, I have more trouble with that: believed by whom or what?

The above criteria for knowledge seem like common sense, and the ancient Greek philosophers had a real knack for encapsulating the common sense view of the world in their philosophy. But common sense is frequently wrong, so let us look at those criteria with a more jaundiced eye. Let us start with the first criteria: it must be justified. How do we justify a belief? From the sophists of ancient Greece, to the postmodernists and the-anything-goes hippies of the 1960s, it has been demonstrated that what can be known for certain is vanishingly small.

----

[1] On the other hand, oligarchy, as any large multinational will tell you, is very profitable.
[2] Also used in business management under the rubric of the Plan-Do-Check-Act cycle.

Renee Descartes (1596 – 1960) argues in the beginning of his DIS-COURSE ON THE METHOD (1673) that all knowledge is subject to doubt: a process called methodological skepticism. To a large extent, he is correct. Then to get to something that is certain he came up with his famous statement: *I think, therefore I am*. For a long time this seemed to me like a sure argument. Hence, I *exist* seemed an incontrovertible fact. I then made the mistake of reading Nietzsche[1] (1844 – 1900). He criticizes the argument as presupposing the existence of *I* and *thinking* among other things. It has also been criticized by a number of other philosophers including Bertrand Russell (1872 – 1970). To quote the latter[2]:

> *Some care is needed in using Descartes' argument. "I think, therefore I am" says rather more than is strictly certain. It might seem as though we are quite sure of being the same person today as we were yesterday, and this is no doubt true in some sense. But the real Self is as hard to arrive at as the real table, and does not seem to have that absolute, convincing certainty that belongs to particular experiences.*

Oh well, back to the drawing board.

The criteria for knowledge, as postulated by Plato, lead to knowledge either not existing or being of the most trivial kind. No belief can be absolutely justified and there is no way to tell for certain if any proposed truth is an incontrovertible fact. So where are we? If there are no incontrovertible facts we must deal with uncertainty. In science we make a virtue of this necessity. We start with observations, but unlike the logical positivists we do not assume they are reality or correspond to any ultimate reality. Thus following Immanuel Kant (1724 – 1804) we distinguish the thing-in-itself from its appearances. All we have access to is the appearances. The thing-in-itself is forever hidden.

But all is not lost. We make models to describe past observations. This is relatively easy to do. We then test our models by making testable predictions for future observations. Models are judged by their

---

[1] Reading Nietzsche is always a mistake. He was a madman.
[2] From: The Problems in Philosophy, page 16.

track record in making correct predictions—the more striking the prediction the better. The standard model of particle physics' prediction of the Higgs[1] boson is a prime example of science at its best. The standard model did not become a fact when the Higgs boson was discovered, rather its standing as a useful[2] model was enhanced. It is the reliance on the track record of successful predictions that is the demarcation criteria for science and I would suggest the hallmark for defining knowledge. Knowledge is, thus, any information that allows us to make successful predictions for future observations. Scientifically derived models and the observations they are based on are, then, our only true knowledge. However, to mistake them for descriptions of the ultimate reality or the thing-in-itself would be folly, not knowledge.

14. GOOD MANAGEMENT IS SCIENCE

Management done properly satisfies Sir Karl Popper's (1902 – 1994) demarcation criteria for science, i.e. using models that make falsifiable or at least testable predictions. That was brought home to me by a book[3] by Douglas Hubbard on risk management where he advocated observationally constrained (falsifiable or testable) models for risk analysis evaluated through Monte Carlo calculations. Hmm, observationally constrained models and Monte Carlo calculations, sounds like a recipe for science.

Let us take a step back. The essence of science is modeling how the universe works and checking the assumptions of the model and its predictions against observations. The predictions must be testable. According to Hubbard, the essence of risk management is modeling processes and checking the assumptions of the model and its predictions against observations. The predictions must be testable. What we

---

[1] To be buzzword compliant, I mention the Higgs boson.
[2] This view of science is consistent with the philosophic tradition know as pragmatism.
[3] THE FAILURE OF RISK MANAGEMENT: WHY IT'S BROKEN AND HOW TO FIX IT by Douglas W. Hubbard (Apr 27, 2009)

are seeing here is a common paradigm for knowledge in which modeling and testing against observation play a key role.

The knowledge paradigm is the same in project management. A project plan, with its resource loaded schedules and other paraphernalia, is a model for how the project is expected to proceed. To monitor a project you check the plan (model) against actuals (a fancy euphemism for observations, where observations may or may not correspond to reality). Again, it reduces back to observationally constrained models and testable predictions. Or to quote International Standards Organization (ISO) 9000 quality management principles: *decisions based on the analysis and evaluation of data and information are more likely to produce desired results.*

The foundations of science and good management practices are tied even closer together. Consider the PDCA (Plan-Do-Check-Act) cycle for process management that is present, either implicitly or explicitly, in essentially all the ISO standards related to management. It was originated by Walter Shewhart (1891 – 1967), an American physicist, engineer and statistician, and popularized by Edwards Deming (1900 – 1993), an American engineer, statistician, professor, author, lecturer and management consultant. Engineers are into everything. The actual idea of the cycle is based on the ideas of Francis Bacon (1561 – 1629) but could equally well be based on the work of Roger Bacon[1] (1214 – 1294). Hence, it should probably be called the Double Bacon Roll (no, that sounds too much like a breakfast food).

But what is this cycle? For science, it is: PLAN an experiment to test a model, DO the experiment, CHECK the model results against the observed results, and ACT to change the model in response to the new information from the check stage or devise more precise tests if the predictions and observations agree. For process management replace experiment with production process[2]. As a result, you have a model for how the production process should work and doing the process allows you to test the model. The *check* stage is where you see if the

---

[1] Roger Bacon described a repeating cycle of observation, hypothesis, and experimentation.
[2] See also the Prologue.

process performed as expected and the *act* stage allows you to improve the process if the model and actuals do not agree. The key point is the *check* step. It is necessary if you are to improve the process; otherwise you do not know what is going wrong or, indeed, even if something is going wrong. It is only possible if the plan makes predictions that are falsifiable or at least testable. Popper would be pleased.

There is another interesting aspect of the ISO 9001 standard. It is based on the idea of processes. A process is defined as an activity that converts inputs into outputs. Well, that sounds rather vague, but the vagueness is an asset, kind of like degrees of freedom in an effective field theory. Define them as you like, but if you choose them incorrectly you will be sorry. The real advantage of effective field theory and the flexible definition of process is that you can study a system at any scale you like. In effective field theory, you study processes that operate at the scale of the atom, the scale of the nucleus, or the scale of the nucleon, and tie them together with a few parameters. Similarly with processes, you can study the whole organization as a process or drill down and look at sub-process at any scale you like, for CERN or TRIUMF that would be down to the last magnet. It would not be useful to go further and study accelerator operations at the nucleon scale. At a given scale different processes are tied together by their inputs and outputs and these are also used to tie processes at different scales.

As a theoretical physicist who has gone over to the dark side and into administration, I find it amusing to see the techniques and approaches from science being borrowed for use in administration, even Monte Carlo calculations. The use of similar techniques in science and administration goes back to the same underlying idea: all true knowledge is obtained through observation and its use to build better *testable* models, whether in science or other walks of life.

15. QUESTIONING THE EXISTENCE OF GOD AND THE GOD PARTICLE

Does God exist? This is one of the oldest questions in philosophy and is still much debated. The debate on the God particle is much more recent but searching for it has cost a large fortune and inspired people's careers. But before we can answer the questions implied in the title, we have to decide what we mean when we say something exists. The approach here follows that of my previous essay that defines knowledge in terms of models that make successful predictions.

Let us start with a simple question: What does it mean when we say a tree exists? The evidence for the existence of trees falls into two categories: direct and indirect. Every autumn, I rake the leaves in my backyard. From this I deduce that the neighbour has a tree. This is indirect evidence. I develop a model that the leaves in my backyard come from a tree in the neighbour's yard. This model is tested by checking the prediction that the leaves are coming from the direction of the neighbour's yard. Observations have confirmed this prediction. Can I then conclude that a tree exists? Probably, but it would be useful to have direct evidence. To obtain this, I look into my neighbour's yard. Yup, there is a tree. But not so fast—what my eye perceives is a series of impressions of light. The brain then uses that input to construct a model of reality and that model includes the tree. The tree we see is so obvious that we frequently forget that it is the result of model construction, subconscious model construction, but model construction nonetheless. The model is further tested when I walk into the tree and hurt myself—another confirmed prediction.

Now consider a slightly more sophisticated example: atoms. The idea of atoms, in some form or other, dates back to ancient India and Greece but the modern idea of atoms dates to John Dalton (1766 – 1844). He used the concept of atoms to explain why elements always interact in the ratios of small whole numbers. This is indirect evidence for the existence of atoms and was enough to convince the chemists but not the physicists of that time. Some like Ernst Mach (1838 – 1916) refused to believe in what they could not see up until the beginning of the last century[1]. But then Albert Einstein's (1879 – 1955) famous 1905 paper[2] on Brownian motion (the motion of small particles suspended in a liquid) convinced even the most recalcitrant

_____

[1] Yes, 1905 was the last century. I AM getting old.
[2] He had more than one famous 1905 paper.

physicists that atoms exist[1]. Einstein showed that Brownian motion could be easily understood as the result of the motion of discrete atoms. This was still indirect evidence but convincing to almost everyone. Atoms were only directly *seen* after the invention of the scanning electron microscope and even then there was model dependence in interpreting the scanning electron microscope results. As with the tree, we claim that atoms exist because, as a shown by Dalton, Einstein and others, they form an essential part of models that have strong track records of successful predictions.

Now on to the God particle. What a name! The God particle has little in common with God but the name does sound good in the title of this essay. Then again, calling it the Higgs boson is not without problems as people other than Peter Higgs[2] (b. 1920) have claimed to have been the first to predict its existence. Back to the main point, why do we say the God particle exists? First there is the indirect evidence. The standard model of particle physics has an enviable record of successful predictions. Indeed, many (most?) particle physicists would be happier if it had had some incorrect predictions.

We could replicate most of the successful predictions of the standard model without the God particle, but only at the expense of making the model much more complicated. Like the recalcitrant physicists of old who rejected the atom, the indirect evidence for the God particle was not good enough for most modern-day particle physicists. Although few actually doubted its existence, like doubting Thomas, they had to see it for themselves. Thus, the Large Hadron Collider (LHC) and its detectors were built and direct evidence was found. Or was it? Would lines on a computer screen have convinced the logical positivists like Ernst Mach? Probably not. But the standard model predicted bumps in the cross-sections and the bumps were found. Given the accumulated evidence and its starring role in the standard model of particle physics, we confidently proclaim that the God particle, like the tree and the atom, exists. But remember, that even for the tree our arguments were model dependent.

---

[1] Except Mach who probably never believed in atoms.
[2] Why do we claim Peter Higgs exists? But, I digress.

Having discussed the God particle what about God? I would apply the same criteria to His/Her/Its existence as for the tree, the atom, or the God particle. As in those cases, the evidence can be direct or indirect. Indirect evidence for God's existence would be, for example, the argument from design attributed to William Paley (1743 – 1805). This argument makes an analogy between the design in nature and the design of a watch. The question is then is this a good analogy? If we adopt the approach of science this reduces to the question: Can the analogy be used to make correct predictions for observations? If it can, the analogy is useful, otherwise it should be discarded. There is also the possibility of direct evidence: Has God or His messengers ever been seen or heard? But as the previous examples show, nothing is ever really seen directly but depends on model construction. As optical illusions illustrate, what is seen is not always what is there. Even doubting Thomas may have been too ready to accept what he had seen. As with the tree, the atom or the God particle, the question comes back to: Does God form an essential part of a model with a track record of successful predictions?

So does God exist? I have outlined the method for answering this question and given examples of the method for trees, atoms and the God particle. Following the accepted pedagogical practice in nuclear physics, I leave the task of answering the question of God's existence as an exercise for you, the reader.

16. ESSENTIALLY, ALL MODELS ARE WRONG, BUT SOME ARE USEFUL

Since model building is the essence of science, this quote has a bit of a bite to it. It is from George E. P. Box (1919 – 2013), who was not only an eminent statistician but also an eminently quotable one. Another quote from him: *One important idea is that science is a means whereby learning is achieved, not by mere theoretical speculation on the one hand, nor by the undirected accumulation of practical facts on the other, but rather by a motivated iteration between theory and practice*[1]. Thus he saw science as an iteration between observation and theory. And what is theory but the building of erroneous, or at least approximate, models?

To amplify that last comment: the main point of my philosophical musings is that science is the building of models for how the universe works; models constrained by observation and tested by their ability to make predictions for new observations, but models nonetheless. In this context, the above quote has significant implications for science. Models, even those of science, are by their very nature simplifications and as such are not one hundred percent accurate. Consider the case of a map. Creating a detailed 1:1 map is not only impractical[2] but even if you had one it would be one hundred percent useless; just try folding a 1:1 scale map of Vancouver. A model with all the complexity of the original does not help us understand the original. Indeed the whole purpose of a model is to eliminate details that are not essential to the problem at hand.

By their very nature, numerical models are always approximate and this is probably what Box had in mind with his statement. One neglects small effects like the gravitational influence of a mosquito. Even as one begins computing, one makes numerical approximations, replacing integrals with sums or vice versa, derivatives with finite differences, etc. However, one wants to control errors and keep them to a minimum. Statistical analysis techniques, such as Box developed, help estimate and control errors.

To a large extent it is self-evident that models are approximate; so what? Again to quote George Box[3]:

> *Since all models are wrong the scientist cannot obtain a "correct" one by excessive elaboration. On the contrary following William of Occam he should seek an economical description of natural phenomena. Just as the ability to devise*

---

[1] Hence the foolishness of talking about theoretical breakthroughs in science. All breakthroughs arise from pondering about observations and observations testing those ponderings.
[2] Not even Google could produce that.
[3] From SCIENCE AND STATISTICS (1976), page 792.

simple but evocative models is the signature of the great sci-entist so overelaboration and overparameterization is often the mark of mediocrity.

What would he have thought of a model with twenty plus parameters, like the standard model of particle physics? His point is a valid one. All measurements have experimental errors. If your fit is perfect you are almost certainly fitting noise. Hence, adding more parameters to get a perfect fit is a fool's errand. But even without experimental error, a large number of parameters frequently indicates something important has been missed. Has something been missed in the standard model of particle physics with its many parameters or is the universe really that complicated?

There is an even more basic reason all models are wrong. This goes back at least as far as Immanuel Kant (1724 – 1804). He made the distinction between observation of an object and the object in itself. One never has direct experience of things, the so-called noumenal world; what one experiences is the phenomenal world as conveyed to us by our senses. What we see is not even what has been recorded by the eye. The mind massages the raw observation into something it can understand; a useful but not necessarily accurate model of the world. Science then continues this process in a systematic manner to construct models to describe observations but not necessarily the underlying reality.

Despite being by definition at least partially wrong, models are frequently useful. The scale model map is useful to tourists trying to find their way around Vancouver or to an army general plotting strategy for his next battle. But, if the maps are too far wrong the tourist will get lost and fall into False Creek and the general will go down in history as a failure. Similarly, the models for weather predictions are useful although they are certainly not a hundred percent accurate. However, they do indicate when it is safe to plan a picnic or cut the hay; provided they are right more than by chance. The standard model of particle physics, despite having many parameters and not including gravity, is a useful description of a wide range of observations. But to return to the main point, all models, even useful ones, are wrong because they are approximations, and not even approximations to reality, but to our observations of that reality. Where does that leave us? Well, let us save the last word for George Box: *Remember that all models are wrong; the practical question is how wrong do they have to be to not be useful.*

17. THE RELATIVITY OF WRONG

Isaac Asimov (1920 – 1992) was a prolific writer of science fiction and popular science books. Many people of my generation had their introduction to science through his writings. While not considered a philosopher of science, one of his articles should be required reading for anyone hoping to understand how science works. The article, with the same title as this essay, first appeared on THE SKEPTICAL IN-QUIRER (p. 35, v. 14 1989) and later in a book of the same name. The following is my take on his point.

To clarify the relativity of wrong concept, consider the value of π. A simple approximation is $\pi = 3$ (I Kings 7:23). This is wrong but by less than 5%. A better approximation is $\pi = 3.14$. The error here is 0.05%. Strictly speaking both values are wrong. However, the second value is less wrong than the first. As a graduate student (in the olden days, as my daughter would say) I used $\pi = 3.141592653589793$ in my computer programs. This is still wrong but much less wrong than the previous approximations. There was no point using a more accurate value of π since the precision of the computer was 15 digits (single precision on a CDC computer). None of these values of π are absolutely correct. That would take an infinite number of digits, so all are wrong. However the initial values are more wrong than the subsequent values. They all are useful in the appropriate context. Hence, the relativity of wrong.

The same logic applies to models. Consider the flat earth model. For the person who never travels farther that 100 km from his birthplace, the flat earth model is quite accurate. The curvature of the earth is too small to be detected. However when the person is a sailor, the question of the shape of the earth takes on more urgency. The flat earth model suggests questions like: Where is the edge of the earth? What will happen if I get too close? For the world traveler, the flat earth model is not sufficient. The spherical earth model is more useful, has greater predictive power and suggests a wider range of questions. Questions like: Does the earth rotate? Does it move around the sun or does the sun move around the earth? But it is a wrong statement that the earth is exactly spherical. Not as wrong as the statement the earth is flat but still wrong. However being not exactly correct does not make it useless. A spherical globe allows a much better understanding of airplane routes than a flat map. But the earth is not a perfect sphere. It is flattened at the poles (a quadrupole deformation). Smaller still is its octapole (pear shaped) deformation. The exact shape of earth will never be measured, as that would require not only an infinite number of digits (like π) but also an infinite number of parameters. It would also be useless. What is needed is a description sufficiently accurate for the purpose it is being used for.

Science is the art of the appropriate approximation. While the flat earth model is usually spoken of with derision, it is still widely used. Flat maps, either in atlases, road maps or Google maps, use the flat earth model (except for my road map where it is a crinkled earth model) as an approximation to the more complicated shape.

Classical mechanics—Newton's law of motion and Maxwell equations of electromagnetism—although superseded by relativity and quantum mechanics, are still useful and taught in university courses. The motion of the earth around the sun is still given by Newton's laws and classical optics still works. However, quantum mechanics has a much wider realm of reliability. It can describe the properties of the atom and the atomic nucleus where classical mechanics fails completely.

Animals reproducing after their kind is the few-generations limit of evolution. Thus, over the time scale of few human generations we do not see new kinds arising. The offspring resemble their parents. Evolution keeps the successes of the previous model; cats do not give birth to dogs, nor monkeys to people, even in evolution. The continuity between animals reproducing after their kind and evolution is not sufficiently appreciated by the foes of evolution and perhaps not by its proponents either.

There is a general trend: new models reduce to the previous model for a restricted range of observations. Ideally the new model would contain all the successes of the old model but this is not always the case. But overall the new model must have more predictive power; otherwise its adoption is a mistake. Thus we have the view of science producing a succession of models, each less wrong than the one it replaces but none 100% correct. We see progress. Science progresses, new models are constructed with greater and greater predictive power. The ultimate aim is to have a model of everything with a strictly limited number of assumptions. This model would describe or predict all possible observations. Quantum indeterminacy suggests that such a model does not exist. However, progress in science is moving closer to this ultimate, probably unreachable, yet still enticing goal.

18. SCIENCE, CAPITALISM AND EVOLUTION

In the June 2011 compilation of the top 500 fastest computers in the world, 91% run some version of the Linux/GNU operating system. This compares with 1.2% running MS Windows. A student, Linus Throvalds, and a ragtag bunch of developers across the internet initially developed the kernel of the LINUX/GNU. It has since become a major industrial enterprise but still has developers spread across the internet. The development process for the kernel has been criticized, most notably by Microsoft, for its lack of road map or central planning. Thorvalds considers this a feature, not a bug, and stated the point very clearly[1]:

> And don't EVER make the mistake that you can design something better than what you get from ruthless massively parallel trial-and-error with a feedback cycle. That's giving your intelligence much too much credit.

At 91% to 1.2%, he probably is onto something (not to mention the android/Linux cell phone market share: android: 36%, MS Windows Phone 7: 1% in mid 2011).

---

[1] In the fa.linux.kernel news group in 2001. In the same post he predicted the demise of Sun Mircosystems.

Okay, so what has that got to do with science, capitalism, or evolution? Quite a bit actually. The thesis of this essay is that all three of these are successful due to the superiority *of ruthless massively parallel trial-and-error with a feedback cycle* to central planning.

We go back to Michael Polanyi (1891 – 1976). In a trip to the Soviet Union in 1936 he was told the distinction between pure and applied science was mistaken, and that in a socialist society all scientific research takes place in accordance with the needs of the latest Five-Year Plan. Polanyi, in reaction, showed science behaves much like a free market in ideas with the corollary that central planning is as destructive in science as in the economy. A typical quote from Polanyi[1]:

> *Any attempt at guiding scientific research towards a purpose other than its own is an attempt to deflect it from the advancement of science. (...) You can kill or mutilate the advance of science, you cannot shape it. For it can advance only by essentially unpredictable steps, pursuing problems of its own, and the practical benefits of these advances will be incidental and hence doubly unpredictable.*

The essential point is *unpredictable.* In the long term, science is too unpredictable to control in any useful manner. We do not know in advance which line of inquiry will lead to breakthroughs or whether the breakthroughs will be for good or evil. If we knew, there would be no need for the research. The same reason holds for the failure of central planning in the economy: the problem is too complex and unpredictable. Central planning only works when the system under consideration is simple and predictable.

So what do we replace central planning with? Back to Thorvalds's *ruthless massively parallel trial-and-error with a feedback cycle.* The massively parallel process permits many different approaches to be explored simultaneously and results are obtained in a timely manner. We do this in science by having different scientists work on different approaches to a given problem. The ruthlessness comes in rejecting or ignoring all the approaches that fail. Most of what is published in

[1] From THE REPUBLIC OF SCIENCE, page 52.

science is ignored and only a few papers have a significant impact. I have seen the statement that, on average, a published paper is read only twice. This means most published papers are never read at all (perhaps not even by all the authors). There is no way to tell in advance which research will fall into the unread category. You try all approaches and see which ones work.

It is similar in a capitalist society. Companies try many different approaches. The ones that work make their owners rich, while the ones that fail go bankrupt. The examples are legendary: IBM moved with the times and, for a while, was almost synonymous with computers (hence the slogan; no one ever got fired for buying IBM). Digital Equipment Corporation (DEC), once a computer heavy weight, faded into oblivion (CEO Ken Olsen: There is no reason for any individual to have a computer in his home.)

The ruthlessness in capitalism comes in by allowing companies to fail. Capitalism breaks down when companies become too big to fail, or with monopolies and oligarchies. Companies that are too big to fail, monopolies, and oligarchies might as well be run by the government since they have lost the attribute (ruthless feedback) that makes the capitalist model work.

Evolution works on the same principle as the capitalist society and Linux development (as noted by Thorvalds). There is no central planning but a massively parallel system with a very ruthless feedback loop: the poorly adapted die. It is ironic that the left attacks capitalism and supports evolution, while right attacks evolution and supports capitalism since both evolution and capitalism depend on one and the same principle: self-organization through ruthless feedback.

For simple systems, central planning does work. Science or economies can be directed but only in the short term. Eventually the Soviet Union collapsed. So will science—if we regulate it too closely.

19. THE ROLE OF PURE RESEARCH

Thomas Edison (1847 – 1931) was a genius. He was also the ultimately practical person devoted to producing inventions with commercial applications. His quote on airships from 1897 is typical: *I am not, however, figuring on inventing an airship. I prefer to devote my time to objects which have some commercial value. At the best, airships would only be toys.* Fortunately the Wright brothers liked playing with toys and indeed the airplane was just a toy for many years after it was first invented. But just ask Boeing, Airbus or even Bombardier if airplanes are still toys. Progress requires both the practical people, like Edison, and the people who play with toys, like the Wright brothers.

Let's pick on Edison again. The practical Edison patented something known as the Edison effect, but did nothing more with it. The effect was this: if a second electrode is put in a light bulb, it is found that an electrical current would flow if the voltage was applied in the right direction. This led to the diode which improved radio reception and in the hands of people, who liked playing with toys, led to the vacuum tube. The vacuum tube is now largely obsolete but began the electronics revolution. Again, we see that progress depends on the people who like playing with toys as well as the people concerned with immediate practical applications. The practical use of an observation, like the Edison effect, is frequently not immediately obvious.

With the light bulb, Edison played a different role. The light bulb is at the end of the chain of discovery. It relies on all the impractical work of people like Michael Faraday (1791 – 1867) and James Maxwell (1831 – 1879), who developed the ideas needed for the practical generation and transmission of electrical power. Without the power grid, which their discoveries made possible, the light bulb would have only been a toy.

The discovery of radium is another example of a pure research project leading to practical results. At one time, radium was used extensively to treat cancer. To quote Madam Marie Curie[1] (1867 – 1934):

We must not forget that when radium was discovered no one knew that it would prove useful in hospitals. The work was

---

[1] The first person to win two Nobel prizes. This quote is from a Lecture at Vassar College, Poughkeepsie, New York (14 May 1921).

one of pure science. And this is a proof that scientific work must not be considered from the point of view of the direct usefulness of it. It must be done for itself, for the beauty of science, and then there is always the chance that a scientific discovery may become like the radium a benefit for humanity.

An even more striking example of how serendipitously science advances technology is the modern computer. It relies on transistors which are very much quantum devices. The early development of quantum mechanics was driven by the study of atomic physics. So, I could just imagine Earnest Rutherford (1871 – 1937), an early experimenter in atomic physics, thinking: *I want to help develop a computing device so I will scatter some alpha particles.* Not bloody likely! The implications of pure research are simply unknowable. However, I doubt the Higgs boson will ever have practical applications. The energy scale is simply too far removed from the everyday scales. Rutherford was probably criticized with the same argument.

But pure research contributes to society in another way. A prime example is the cyclotron. It was invented in 1932 for use in the esoteric study of nuclear physics. Initially, cyclotrons were only in top physics departments and laboratories. Now they are in the basements of many hospitals were they are used to make rare isotopes for medical imaging and treatment. The techniques developed for pure research frequently find their way into practical use. While standing on the moon did not produce any real benefits to mankind, the technology developed in the enterprise did; hence the term: *space age technology.*

Of course, I cannot leave this topic without bring up the World Wide Web. The initial development was done at CERN in support of particle physics. I remember a colleague getting all excited about this new software development, but initially it was something only a geek like her could love. The links were denoted by numbers that had to be typed in, no clicking-on-links. Then the National Center for Supercomputing Applications (NCSA) at the University of Illinois Urbana-Champaign developed a browser, Mosaic, with a graphical interface and embedded pictures. This browser was released in 1993 and looks much like any browser today. The rest is history. But, two other things were needed to make the *World Wide Web* a hit. The first was computers (those things that were developed from Rutherford scattering alpha particles) with sufficient capabilities to run the more powerful browsers and, of course, the internet itself. The internet was initially just an academic network but the World Wide Web provided the impetus to drive it into most homes. Here again we see a combination of efforts: academic at CERN and NCSA and commercial at providing the internet and the more sophisticated browsers that followed Mosaic.

Thus we see pure research providing the raw material for technological development. The raw material is either the models, like quantum mechanics, or the inventions, like cyclotrons. These are then used by practical men like Edison to generate useful technology. However, there is also a cultural component: satisfying our curiosity. While the spinoffs may be the main reason politicians and the taxpayers support pure science, it is not the motivation driving the scientists who work in pure science. In my own case, I went into physics to understand how the universe works. To a large extent that desire has been fulfilled, not so much by my own efforts but by learning what others have discovered. More generally the driving force in pure science is curiosity about how the universe works and the joy of discovery. Like Christopher Columbus (1451 – 1506), Robert Scott (1868 – 1912) or Captain James Kirk (b. 2233), pure scientists are exploring new worlds and going where no man, or woman, has gone before.

20. WHAT'S BEST: EXPERIENCE OR PURE THOUGHT?

Pure logical thinking can give us no knowledge whatsoever of the world of experience; all knowledge about reality begins with experience and terminates in it. Before you accuse me of scientism[1] let me point out that the previous sentence is a direct quote from Albert Einstein (1879 – 1955). Henri Poincaré (1854 – 1912) agreed: *Experiment is the sole source of truth. It alone can teach us something new;*

---

[1] I will defend scientism in another essay.

51

it alone can give certainty. These are two points that cannot be questioned. Wow, tell us what you really think, don't hold anything back. I might note that both Einstein and Poincaré were theorists.

Two points that cannot be questioned? They certainly have been questioned—all the way back to Plato. The so-called continental rationalists, people like René Descartes (1596 – 1650, *I think therefore I am*), Baruch de Spinoza (1632 – 1677, a pantheist), Gottfried Leibnitz (1646 – 1716, Newton's rival in inventing calculus), and Immanuel Kant (1724 – 1804, of synthetic a priori knowledge), based their epistemology on pure thought. Take Descartes—he developed an extensive physics based on pure thought with planetary motion due to vortices. Never heard of it? Shows the folly of trusting pure thought; it sunk without a trace under Newton's empiricism. Kant developed the idea of synthetic a priori knowledge; knowledge that came from pure thought and not observation. Unfortunately, his examples— Euclidean geometry and Newton's laws, turned out not to be true, at least not as the ultimate models of physical reality. Oops. At the risk (or pleasure[1]) of offending some people, I add proofs of God's existence or non-existence to the list of failed attempts to obtain knowledge by pure thought.

But the lure of obtaining knowledge by pure thought is tempting: certainty AND a free lunch. No need to talk to those annoying experimentalists who keep shooting my theories down. One of the people who succumbed to the temptation of pure thought was David Hume (1711 – 1776). This is more surprising since he was a phenomenologist to the core. But he did not like miracles and tried to eliminate miracles by arguments based on pure thought. Well, if Einstein and Poincaré are correct, Hume is wrong. And in my opinion, wrong he is. It all hangs on the question: What is a miracle?

Before we address that, let us tackle a simpler question: What is the distinction between natural and supernatural? Consider thunder and lightning. The Vikings believed that thunder was the noise made by the wheels of Thor's chariot being pulled across the heavens by goats. This view was reinforced by the sparks made from the chariot wheels

---

[1] I have a firm policy of never ACCIDENTALLY offending anyone.

hitting rocks; sparks otherwise known as lightning. Groves of trees—which are the prime target for lightning—became sacred. Today we have a more prosaic view of thunder and lightning—just electromagnetism. The phenomena have not changed but the meaning has, in one thunder is supernatural, in the other natural. Observations are given meaning based on the model or paradigm (Kuhn's nomenclature) used to describe them.

We have an apparent collision between Kuhn on the one hand and Einstein and Poincaré on the other: models giving meaning vs observation being paramount. But it is more in appearance than reality. Observations are used to help build and constrain models, while the models then give meaning to the observations. This is self-consistent, not circular. The wiggles in the data seen recently at the LHC are only meaningful within the context of a model for high-energy physics and the detector. Was finding the Higgs boson a miracle? Probably not[1]. But super-symmetry... that is another matter.

More seriously: What is a miracle? According to Hume, a miracle is something that violates the laws of nature. That would be fine if we had a definitive list of the laws of nature. We don't. We have, at best, something that may approximate them, something obtained by observations. If miracles occur, they would be observed and therefore built into in the observationally derived laws, rendering Hume's definition meaningless. Rather, we define miracle as something that is supernatural. By the argument above, it then depends on the model: the model that is constrained by observation. In the end, the existence of miracles and all other questions of how the universe operates have to be settled empirically by observations and the models built on them. The medium may be the message, but the meaning is in the (observationally-constrained) model.

---

[1] Written before the Higgs boson was found. Perhaps finding it was a miracle.

Sometime ago, one of my acquaintances told me that his son had shown evolution was not a fact. This set me to thinking furiously, as Hercule Poirot would say. No, not about the validity of evolution. I am not a biologist and besides evolution is one of the best-supported scientific models of all time, like classical mechanics in 1900. Ah, perhaps that is not the best example, but you know what I mean.

Rather, it set me to thinking furiously about how to explain what science is to the uninitiated. After hearing this statement, during my daily commute to and from work[1], I would think about what science is, how it works, and how to explain it to the general public. The result is given in my article in PHYSICS IN CANADA (2007, Vol. 63, p. 7) and in this series of essays. In the case of my acquaintances mentioned above, I suspect it would fall on deaf ears. After all, they listen to a faux news network.

Evolution is not a fact. This statement strikes me as peculiar; like the statement: *The number three is not green.* Evolution is a model of how species arise. It makes about as much sense to ask if a scientific model (theory, law, or paradigm) is a fact as it does to ask if a model boat is a fact. The question is poorly posed to the point of being meaningless. (There is no such thing as a stupid question but if there were that would be it.) But one hears such questions all the time: Is global warming a fact? Is it a fact the universe 13.4 billion years old? Is it a fact your mother wears army boots? (Well, maybe not the last.) It is also common for scientists to get that deer-in-the-headlights look when asked such questions. So what is going on?

The answer came to me in a lecture by Carl Wieman (b. 1951, physics Nobel Laureate and science educator). He pointed out that, in any field, the novice and expert view the field differently. Whether it is chess or biology, the novice tends to view the field as a catalogue of facts (an idea that goes back at least to Aristotle), and the important point is what are these facts. Hence the question: Is that a fact? i.e. Should it be in the catalogue? The expert on the other hand sees pat-

---

[1] Fortunately, I did not run over anyone.

terns, relationships and organization but has no catalogue of true statements. There is no need to remember individual pieces of information because they are inherent in the patterns and can be reconstructed from them. For the expert, it is the patterns and relationships that are important. In science, these patterns are called theories (although I prefer the term model). Thus, we have two incommensurate views (or paradigms) of knowledge: a catalogue of largely unrelated facts or a coherent framework (a model describing observations).

When the novice or layman asks if it is a fact, he wants to know if the statement is to be included in the catalogue. He knows nothing of theories as general organizing principles since they are not part of his understanding of science. As an example of this, see the Answer to the Complaint in the Kitzmiller v. Dover Area School District court case on the role of evolution in the classroom where it is stated: *Defendants deny that the term theory, as used in science, has a distinct meaning and does not suggest uncertainty, doubt or speculation.* The first response to this is: Can the defendants not read a dictionary? (Check the definition of theory in just about any dictionary). Are they actually *that* stupid? But that is not the problem: In their understanding of science as a collection of facts, there is no place for general organizing principles.

Theirs is a Victorian (or Newtonian) view of science:

> *Scientists, it seemed clear, began with careful observations, cautiously proceeded to a tentative hypothesis, progressed to more secure but still provisional theories, and only in the end achieved, after a long process of verification, the security of permanent laws1.*

This view of science has been superseded, just as quantum mechanics superseded classical mechanics. Contemporary scientists, the experts in this case, have in their model of science no place for the catalogue of true statements or permanent laws. They have observations (whose interpretation is model dependent) and observationally constrained models, otherwise known as theories. These are theories, not in the

---

[1] A PILGRIMAGE TO POPPER, Adam Gopnik, The New Yorker 2002-04-01

sense of conjectures, but in the sense of general organizing principles: the special theory of relativity, quantum theory, or the theory of evolution. The scientist and the general public talk past each other using the same words, but meaning different things: two cultures divided by a common language[1].

22. THE MEANING IS IN THE MODEL: EVEN ON VACATION

Here I sit on vacation, looking out the window. I see the grass reaching to the cliff edge, a Douglas fir tree, the blue water, a small island, and in the distance the mountains on the B.C. mainland. All in all, a pretty picture. Oh, I forgot to mention the crane doing some work shoring up the cliff face and standing in the middle of the picture. There's always something. And the wifi is not working and the daughter wants to go do something.

But what do I actually see? The eyes detect some electromagnetic radiation lumped into three arbitrary groupings and from this, constructs a model of the surroundings which it presents to the conscious mind. The three dimensional layout, grass, trees and water are all constructions of the unconscious mind. The human perceptual apparatus creates a model that we mistake for reality. The grass and water are automatic. The Douglas fir tree requires a little conscious effort to identify based on previous knowledge of trees. That the land out there is an island and the far land is the mainland relies on my knowledge of geography and which way I am looking. The knowledge of the cliff relies on my looking around last night and the assumption that things do not change capriciously over time. Models within models. All constructed by the human mind, most unconsciously, a few consciously.

Now, the reductionist would claim the model is seriously flawed: no atoms, electrons, atomic nuclei, or quarks. String theorists would even claim the number of dimensions is wrong. But it is a very good model for scales from a millimeter to kilometers. It was honed by

---

[1] Paraphrase of a common expression of uncertain origin.

evolution and gives meaning to our surroundings; meanings like: *that tawny colored blob over there is a cougar[1]! Run!* (Well, running from a cougar is not a good idea.)

In everyday life, as in science, the meaning is in the model (and the context), not in the raw observations. Examples abound: consider the idea of naturalism, discussed in the essay on *Pure Reason*. Both in philosophy (naturalism) and in marketing (natural products), the meaning comes from the model. In marketing, natural seems to mean using techniques developed before 1900 or thereabouts. And we laugh at the Amish. But I digress.

Cause and effect is also something that comes out of the models used to describe observations, not in the observations themselves. As the statisticians say: correlations do not imply causation. The sky lighting up in the morning does not cause the sun to rise, despite the invariant correlation with the light preceding the rising of the sun. I have seen a philosophy paper wax elegant for pages, trying to determine if a fence post causes the shadow or the shadow the post, without building an explicit model. It may be possible, but why bother. That is not how science or the mind works. Cause and effect come out the model.

Similarly, it is the models that provide explanations. While the prime goal of science is to construct models that describe observations, these models frequently provide explanations. Why do the planets have retrograde motion? They and the earth circle the sun (or they move on epicycles, take your pick). Why do people get the common cold? Viruses. Why was your high school science teacher such a dork? Well, no scientific model can provide an explanation for that.

While the meaning given to observations is model dependent, contrary to Kuhn, the observations themselves tend not to be. The moving lights in the night sky (not airplanes, the planets) have been described using various models, from Gods to objects orbiting the sun. But the observations remain the same: lights moving about in the night sky. The Ptolemaic and Copernican systems may be different paradigms, as are Aristotelian physics and quantum mechanics, but they all agree

---

[1] The cougar showed up a few days after we left.

on the observations regarding the lights moving in the sky. Even string theorist would have to agree.

Thus science has a firm ground to stand on: observation. More specifically, largely *model independent* observation (Kuhn be damned). Otherwise we would be building on sand. But speaking of sand, I must go and see the sand castle competition. Magnificent creations, but built with sand on sand. Perhaps building on sand is not always a bad idea.

23. METHODOLOGICAL NATURALISM

Pierre-Simon, marquis de Laplace (1749 – 1827) was one of the great French mathematical physicists. In math, his fame is shown by the number of mathematical objects named after him: Laplace's equation, Laplace transforms, the Laplacian, etc. In physics, he was the first to show that planetary orbits are stable and he developed a model—the nebular model—to account for how the solar system formed. In modified form, the nebular model is still accepted. In spite of these important contributions, he was also very much a lackey, being very careful to keep on the right side of all the right people. During the French revolution, that might have been just good survival strategy. After all, he served successive French governments and, unlike Antoine Lavoisier (1743 – 1794), kept his head.

Laplace presented his definitive work on the properties of the solar system to Napoleon (1769 –1821). Napoleon, liking to embarrass people, asked Laplace if it was true that there was no mention of the solar system's Creator (i.e. God) in his *opus magnus*. Laplace, on this occasion at least, was not obsequious and replied, *I had no need of that hypothesis*. This is essentially the simplicity argument discussed in a previous essay, but stated very crisply and succinctly.

Laplace was not just a whistlin' Dixie. Newton had needed that hypothesis, God, to make the solar system work. Newton believed that the planetary orbits were unstable and unless God intervened periodically, the planets would wander off into space. Newton had not done a complete mathematical analysis. Laplace rectified the problem.

Newton also had no model for the origin of the solar system. Laplace eliminated these two gaps that Newton had God fill.

Back to Napoleon—he told[1] Joseph Lagrange (1736 – 1813), another of the great French mathematicians/physicists, Laplace's comment about no need for the God hypothesis. Lagrange's reply was, *Ah, it is a fine hypothesis; it explains many things.* Laplace's apocryphal reply was, *This hypothesis, Sir, explains in fact everything, but does not permit to predict anything. As a scholar, I must provide you with works permitting predictions.* This is the ultimate insult in science: it explains everything but predicts nothing. Explanations are a dime a dozen; if you want explanations, read Rudyard Kipling's (1865 – 1936) JUST SO STORIES. Now, there are some fine explanations. I particularly like THE CAT THAT WALKED BY HIMSELF.

Laplace's argument, *I had no need of that hypothesis,* is still being used today. Hawking and Mlodinow in their book, THE GRAND DESIGN, created a stir by claiming God did not exist. But their argument was just Laplace's pushed back from the beginning of the solar system to the beginning of the universe: they had no need of that hypothesis. Whether their physics is correct or not is still an open question. It is not clear that string theory has gotten past the *it explains everything but predicts nothing* stage.

An alternate approach to understanding God's absence in scientific models is methodological naturalism. The term seems to have been coined by the philosopher Paul de Vries, then at Wheaton College, who introduced it at a conference in 1983 and published it in the CHRISTIAN SCHOLAR'S REVIEW. It has since then become a standard definition of science, even playing a significant role in court cases, most notably the case[2] in Dover Pennsylvania on teaching creationism in public schools. The judge mentioned methodological naturalism prominently in his ruling.

---

[1] The latter parts of this story are increasing apocryphal, but it is a good story.

[2] United States District Court for the Middle District Of Pennsylvania, Tammy Kitzmiller, et al. v. Dover Area School District; et al.

Methodological naturalism, as a definition of the scientific method, is rather ill defined except for its main idea, namely that science, explicitly, by fiat, and with malice aforethought, rejects God, gods, and the supernatural from all its considerations. There is frequently an implicit secondary idea that science is about finding explanations but only natural ones, of course. Both ideas are inconsistent with what science actually is: building models constrained ONLY by observation and parsimony. (See above and a previous essay for my opinion of the role of explanations in science.)

However, methodological naturalism is a very convenient hypothesis. It avoids awkward questions about the relation between science and religion. By inserting naturalism into the very definition of science, methodological naturalism, if valid, would create a firewall between science and religion. This would both protect religion from science and scientists from the religious. Considering the violence done in the name of religion, the latter may be more important, but the former was probably part of the original intent. However, I suspect the main motivation was to explain why God and the supernatural are absent from science. But Laplace gave the real reason for God's absence: parsimony—there is no need of that hypothesis. There are probably also very good theological reasons for that absence but that is outside the scope of science and this essay.

Methodological naturalism confuses the input with the output. To the extent science is naturalistic; it is an output of the scientific method, not part of the definition. Excluding anything by fiat is poor methodology. But once one realizes that historically God and the supernatural have been eliminated from science, not by fiat, but by Laplace's criteria, methodological naturalism becomes redundant; an ad hoc solution to an already solved problem.

24. IS SCIENCE OBJECTIVE OR CULTURAL?

Franz Boas (1858 – 1942) was another scientist trained as a physicist who made a name for himself in another field, in this case, anthropology. He was the founder of modern anthropology and brought to the field the methodology of the natural sciences; the idea one should formulate theories and conclusions only *after* thorough and rigorous collection and examination of hard evidence. In cultural anthropology, he established the contextualist approach to culture, cultural relativism. Culture can be thought of as the paradigm that gives context and meaning to social interactions. Cultural relativism recognizes that comparing cultures has the same incommensurability problems as comparing other paradigms. The same words (actions) have different meanings depending on the paradigm (culture). To see how this different meaning works in practice, consider headgear. The rules on what is acceptable head covering is cultural and religious and varies over time and place. What one group considers good and proper, another considers inappropriate. At one time, the English considered the Irish uncouth because they doffed their hats to people they met on the street. The horror of it. At another time, no self-respecting woman would appear in church without a hat (following Saint Paul's instructions), but now Muslim women are sometimes criticized for covering their heads. The Canadian Legion considered it an insult to the Queen not to remove head covering in Legion halls. At least in the case of Sikhs, the Queen did not agree. What one culture praises, another condemns.

Cultural relativism had two main tenets: 1) all people are civilized and 2) there are no higher and lower cultures. This gave a much-needed antidote to the evolutionist idea that preceded it; the idea of the innate and absolute superiority of the western culture, since it was considered more evolved than other cultures. Western culture was then used as the hallmark against which other cultures were judged. In England, it was the *White Man's Burden* (now mostly dead) and in the USA, *American Exceptionalism* (not mostly dead). To understand exceptionalism, think of Raskolnikov in CRIME AND PUNISHMENT— he thought himself exceptional so he did not have to follow normal behavioural conventions. Another of Dostoyevsky's works states that most people are not sufficiently intelligent to realize they are not exceptional. Be that as it may, the only truly exceptional people are Nova Scotia-born physicists[1]. Hmm, perhaps I should not have juxtaposed that with Dostoyevsky's statement. However, it seems all peo-

---

[1] It is entirely coincidental that this describes the author.

ple like to think that their own particular group, culture, or religion is exceptional, so why not Nova Scotian physicists? Cultural relativism is a direct attack on this common idea that one's own group is exceptional or superior. It instead says that all cultures should be evaluated and judged on their own merits, not against the standards of another culture.

Unfortunately, the idea of cultures being self-contained and statements being valid only within a given culture has been extended too far, to exclude all cross-cultural statements. But in the context of science, what does it mean? In some cultures, does the sun rise in the west and set in the east? Or is *the sun rises in the east*, a cross-culturally valid statement? Can we solve the energy crisis by finding a culture where the second law of thermodynamics does not hold— *Build your perpetual motion machines in Lower Slobbovia!*—or is the second law cross-cultural? I mean, in Australia the swans are black, the sun is in the north and they play Aussie rules football. But as far as I know, all the models of science are equally valid there (except perhaps on the football pitch). It is only in Douglas Adams's (1952 – 2001) imagination that we have bistro mathematics. What particles will be found at CERN or Fermilab depends on the nature of the accelerators and detectors, not the culture at the two labs. Trying to change the culture by bringing in mystics or other counter-culture people will not result in finding different types of particles. It has never been observed to work that way.

Now, the supporters of relativism (or its double cousin postmodernism) will complain that the examples I have given are too simple. But a general rule must apply to simple cases, as well as the complex ones where the very complexity makes it hard to see what is happening. If you want me to believe that the model of germs causing disease is only cultural, you must first explain why the model that the sun rises in the east is only cultural. They both result from the same method. In the cold fusion debate, I heard the statement: *if it wasn't for those damn physicists we would have an infinite supply of energy.* If the physicists had not debunked cold fusion, it would still be happening and we would have cold fusion powered Hondas (why not, if all statements are relative, perhaps science is different in Japan). Unfortunately, scientists do not make the laws, only discover them. Culture could not make Lysenkoism[1] valid, even in Stalinist Russia. In the same way that raw observations are valid across scientific paradigms, scientific models are valid (or, as in the case of Lysenkoism, invalid) across cultures or cultural paradigms.

And yet—to balance this argument out—culture and context does play a role in how the results of science are expressed and in who discovers them. A pessimist would be more apt to discover the second law of thermodynamics than an optimist. In England in the 17th century they discovered laws—Hooke's Law, Boyle's Law—but now such regularities are just rules–the OZI rule[2] for example. Thus, how things are expressed changes with the fashion, but the ideas behind them stay the same. If there is an enduring cultural influence in science, it is the culture of mathematics. As Henri Poincaré (1854 – 1912) said[3], *But what we call objective reality…can only be the harmony expressed by mathematical laws. It is this harmony then which is the sole objective reality, the only truth we can obtain.*[4]

25. IN DEFENSE OF SCIENTISM

In the olden days, when everything was in black and white (as my daughter once thought), I attended the Musquodoboit Rural High School. There, I had a number of very good teachers (plus a few horrible ones). One of the good teachers was in Grade 12 English. As good teachers sometimes do, he occasionally got off topic. One class was particularly airy-fairy (as we said in those days) and towards the end he said (complete with a diagram on the blackboard) that we, the students, were over there and when he was our age he was over there too but now he was over here and wanted us to be over here as well.

---

[1] The biological inheritance principle which Trofim Lysenko subscribed to, and which derives from theories of the heritability of acquired characteristics. It contributed to the collapse of Soviet agriculture.
[2] The Okubo, Zweig, Iizuka rule on the decay of excited nucleons and other hadrons. It contributed to the acceptance of the quark model.
[3] Quote from THE FOUNDATIONS OF SCIENCE (THE VALUE OF SCIENCE), published in 1913, page 13.
[4] This idea has been developed into what is known as structural realism.

So I put up my hand and asked what I thought was the obvious question: *How do you know you are going in the right direction?*

It is a question not just for English teachers, but also for everyone: politicians, religious leaders, alternate medicine advocates, car mechanics, and, last but not least, scientists. How do you know you are going in the right direction? The answer to that question is the basis of scientism and the faith in the power of science. What criteria do we have to answer the question? Relatively few:

1. Certainty of conviction? Many different religions have their martyrs. If conviction gave certainty, only the one true religion would have martyrs.
2. Pure thought? Even Descartes failed. No one else has been any more successful. Make one mistake and the whole thing collapses.
3. Divine revelations and spiritual insights? There are conflicting claims to divine revelation. How do you choose between them? Maybe in the attempt to avoid the Christian hell you will end up reincarnated as a worm, or vice versa. I will leave further discussion of this point to the theologians.
4. Innate knowledge? In humans, this does not seem to extend much beyond the fear of heights, spiders, and snakes.
5. Observation? In the end, that is what it all comes down to. Even Deuteronomy says observation and falsification are the way to detect false prophets (Deuteronomy 18:21 – 22).

Now science is built around observation. But even before science started to become mainstream (which I would date to 1610 and Galileo), observation was the bedrock of survival even if Greek philosophers failed to appreciate it. Folk wisdom relies on observation and experience. The scientific method has just systematized the knowledge extraction process: building models and refining them, making definite predictions, testing against observation, controlling errors, and then repeating the process[1]. Using observation as a filter is the first way scientists know they are going in the right direction.

---

[1] The basic idea goes back at least to Roger Bacon (c. 1214 – 1292) but was largely ignored until Galileo, Kepler and Newton.

The success of science is the second reason scientists know they are going in the right direction. Successful at what, you might ask? Certainly not at answering the ultimate question about life, the universe, and everything (see the *next* essay), but at increasing our understanding of how the universe works, at providing the foundation for technology, and of building and maintaining the consensuses that allow the previous two to proceed.

Our understanding of how the universe works has increased enormously: from a flat fixed earth to an expanding cosmos; from earth, air, fire and water to quarks and electrons; from *mal air* causing disease to germs and dietary deficiencies; from killing cats to prevent the plague, to good sanitation. The scientific insights have seeped into the very soul of our culture to the extent that we cannot even imagine what the prescientific world view was like: demons and witchcraft, sacrificing the king to ensure good crops, etc.

Now to technology: science turns money into knowledge and technology turns knowledge into money. That may be crass, but it is fundamentally correct. Science provides the information on how the universe works that technology uses to build useful widgets. Science is useful for technology since it allows us to make accurate predictions, predictions like press this button on the remote and the TV turns on (unless the spouse has unplugged it—nah, that would never happen). Every technological device is a monument to the power of science. The common cell phone depends on the validity of the predictions of classical mechanics, quantum mechanics, special relativity, and even general relativity.

But the real success the scientific method has is in building and maintaining consensus in the scientific community. The core findings of science have widespread support and change only when driven by new observations (experimental results). Essentially all physicists since the time of Newton agree that his model correctly describes planetary motion (except for a small correction for Mercury).

There is, of course, always a fringe of people that disagree about everything, and in the leading edge of science, there is no consensus, but for the most part, the scientific method is successful at driving consensus. Contrast that with religion where denominations multiply and split. There are an estimated 38,000 Christian denominations but there is only one physics and only one scientific method.

Yet, some philosophers of science have the cause and effect backward and see science as primarily a social endeavor where social consensus drives the field. Anyone who has ever worked with scientists knows that is not true. A more independent, competitive, and compulsive group of beings has never existed (well, except maybe cats). But observation and parsimony are hard taskmasters and can keep even scientists in line.

So, science has a clear method to determine that it is going in the right direction and the results to show that the method is working. But what about other fields? How do they know if they are going in the right direction? Maybe, just maybe, we can get a hint from my English teacher's response to the question. His answer was: *You are what I am fighting against*. Then again, maybe not.

26. LET THE MYSTERY BE

The Limits of Science

Many minds, great, mediocre, and small, have pondered from time immemorial the ultimate nature of the universe. They were all searching for the same thing: the answer to the ultimate question of life, the universe, and everything. Naturally, being people, and having no real criteria to decide on the correct answer, they came up with a collection of contradictory answers, including:

1. Materialism: The idea is that what you see is what you get. There is no man behind the curtain manipulating things. In this view, the mind and consciousness arise from the material brain.
2. Idealism: Largely the converse of materialism. Here the mind is fundamental and material objects only exist in the mind.
3. Solipsism: An extreme form of idealism that says that all that exists is my mind. You are out of luck. Or vice versa. This one appeals to me since it makes me the center of the universe.

4. Deism: Materialism with an Omphalic twist. God or gods created the universe and then took an extended coffee break. This tended to be the default view of intellectuals in the age of enlightenment, most notably Thomas Paine (1737 – 1809, see THE AGE OF REASON).
5. Theism: God or gods created the universe and stayed around to interact with their creation. This is typical of western religions—Greek, Roman, Germanic, and the Mosaic religions.
6. Orwellianism: Reality is what the Party says it is. The idea that there is reality apart from what the Party says is a pernicious superstition. This is the extreme case of ideology trumping everything else.
7. Project Management: Only what has been documented exists. I am not sure if this is a subset of Idealism or Orwellianism.
8. 42: If you do not understand this answer you do not know what the question really is (Deep Thought from the HITCH HIKERS GIDE TO THE GALAXY by Douglas Adams (1952 – 2001)).

Number 8 we can ignore (sorry Douglas Adams fans), although the real problem probably is that we do not understand the question. Number 6 is a chilling reminder of what can happen when ideology rules. Number 7 is for my engineering friends (or is that acquaintances). For all the rest, the scientific method, as a method, is agnostic.

Contrary to popular opinion, the scientific method does not assume materialism, realism, or any other -ism. All one needs to carry out science are observations that can be used to construct and test models. Whether the observations are the result of a material world impinging on the mind through the senses, or purely illusions of the mind as in solipsism, does not really matter. The relationship between the models and reality is different for each of the options one through five, but observation cannot discriminate between them. At best, all observation can do is force ever more creative uses of the Duhem-Quine thesis. No matter how materialistic the universe may appear, there is always a place for God to hide; even if a being with vast knowledge and power showed up, there would be no way to prove he was God and not just a being from some highly advanced civilization.

Scientific models depend not just on observation, but also on simplicity—which is the only antidote to the Duhem-Quine thesis. Combining observation and simplicity, the current models of science tend strongly towards materialism. But this could change the next time science lurches in a new direction. Indeed, some claim that this has already happened with quantum mechanics. The measurement process in quantum mechanics is taken by some, possibly misguided souls, to indicate that consciousness has a vital role to play, hence tilting science towards idealism. Quantum mechanics and idealism may be no farther apart than classical mechanics and materialism.

While simplicity is an essential ingredient in constructing our scientific models, can we actually use it as a guide to reality, itself? Perhaps reality is not as simple as our models assume (note the word assume) and the world is only 6,000 years old as Philip Gosse (1810 – 1888) suggested. The lack of God or gods in our scientific models may be only a symptom of the failure of the simplicity assumption. The mind of God, if he exists, is unfathomable, so there is no guarantee that he would respect simplicity.

How do the models of science relate to the ultimate reality? That is unknown and unknowable. In classical mechanics, the particle trajectories and three-dimensional Euclidean geometry were assumed to be real. The first was destroyed by quantum mechanics, the second by relativity. There is no reason to assume the underpinnings and constructs of any current model will not be similarly undermined by future paradigm shifts. Independent of all that, our models stand, giving us the only useful knowledge available for how the universe works. As Niels Bohr stated, *It is wrong to think that the task of physics is to find out how nature is. Physics concerns what we can say about nature.* As for the ultimate nature of reality, I will follow the suggestion of Iris DeMent (b. 1961) in the song LET THE MYSTERY BE:

> *But no one knows for certain*
> *And so it's all the same to me*
> *I think I'll just let the mystery be.*

The blind men and the elephant is a common tale in India with many variants. Below is the one from John Saxe:

The Blind Men and the Elephant
John Godfrey Saxe
(1816 – 1887)

I.

It was six men of Indostan
To learning much inclined,
Who went to see the Elephant
(Though all of them were blind),
That each by observation
Might satisfy his mind.

II.

The First approached the Elephant,
And happening to fall
Against his broad and sturdy side,
At once began to bawl:
"God bless me!-but the Elephant
Is very like a wall!"

III.

The Second, feeling of the tusk,
Cried: "Ho!-what have we here
So very round and smooth and sharp?
To me't is mighty clear
This wonder of an Elephant
Is very like a spear!"

IV.

The Third approached the animal,
And happening to take
The squirming trunk within his hands,
Thus boldly up and spake:
"I see," quoth he, "the Elephant
Is very like a snake!"

V.

The Fourth reached out his eager hand,
 And felt about the knee.
"What most this wondrous beast is like
 Is mighty plain," quoth he;
"'Tis clear enough the Elephant
 Is very like a tree!"

VI

The Fifth, who chanced to touch the ear,
 Said: "E'en the blindest man
Can tell what this resembles most;
 Deny the fact who can,
This marvel of an Elephant
 Is very like a fan!"

VII.

The Sixth no sooner had begun
 About the beast to grope,
Than, seizing on the swinging tail
 That fell within his scope,
"I see," quoth he, "the Elephant
 Is very like a rope!"

VIII.

And so these men of Indostan
 Disputed loud and long,
Each in his own opinion
 Exceeding stiff and strong,
Though each was partly in the right,
 And all were in the wrong!

MORAL.

So, oft in theologic wars
 The disputants, I ween,
Rail on in utter ignorance
 Of what each other mean,
And prate about an Elephant
 Not one of them has seen

In many ways it is an excellent metaphor for science and demonstrates how science works. The blind men want to learn about the elephant so they make observations. Good so far—in science, observations rule. They then make models to describe what they have seen, or in this case, felt. Again, good scientific methodology: the tusk is indeed well modeled as a spear. Since each has chanced upon a different part of the elephant, the models differ. This is not a problem. In science we always model different parts of reality differently; we do not use the standard model of particle physics to describe planetary orbits or cell division. But the blind men then make the classic error and assume theirs is the global model that describes the entire elephant. The scientific method would have required them to make predictions based on their models and test them against additional observations. This would have revealed the problem—and the elephant. Instead, they argued. This, of course, bears no resemblance to real scientists who never argue heatedly on the basis of partial data. Nope, never, has not happened. (Okay, stop thinking of your colleague.)

The moral is also interesting (not just for *theologic wars* but also scientific ones): *Rail on in utter ignorance / of what each other mean.* This seems like a foreshadowing of Kuhn's incommensurability of paradigms. Their ideas and frames of reference are so different they cannot understand each other. *And prate about an Elephant / Not one of them has seen!* Here we come to the nub. To a large extent science is precisely this; prating about an elephant we have not seen. We can learn a lot about the elephant we have not seen by making repeated observations of the beast and learning how the various models fit together and interlock with each other. We may never, by feel or even by sight, be able to construct the ultimate model of the elephant but we can obtain a lot of useful information and construct quite accurate models of at least some aspects of the elephant. The trick is not to make the mistake of assuming one's partial model is the whole truth. This mistake has been made repeatedly: Newton's laws of motion, Maxwell's equation, the fixed continents, etc. There is now a talk of a theory of everything (TOE); the same mistake being made again.

The blind men's error has been common in the philosophy of science as well. Different models for the scientific method capture different aspects of the problem: induction, hypothesis, logical positivism (verification), paradigms, falsification, and even no method. The philosophers do indeed resemble the blind men disputing loud and long. The real task is to see how the various models fit together to give a coherent whole. Like the models of the elephant, the different approaches to the scientific method are not so much wrong as incomplete. The approach I advocate of models competing against each other in making successful predictions is an attempt at a more unified approach; one can see how the various precursor models are aspects of it. But as always, this will probably be just one aspect of a more complete approach that can never be completely known and will always be debated. But for most practical purposes, the scientific method can be, and probably is known well enough. Blind men, even those with a philosophical bent, can learn a lot about an elephant by groping in the dark.

28. REDUCTIONISM VERSUS EMERGENCE

A foreigner was visiting Japan and noticed people playing a game he had not seen before. They had black stones and white stones that they alternately placed on a playing board. He wanted to learn how to play the game but since he did not speak Japanese he decided to learn about the game by watching it being played. (This was before the Internet.) He discovered that the goal of the game seemed to be to surround one's opponent's stones and thus capture them. He noticed that the initial moves seemed to follow a well-defined path but he was not sure if this was from the rules of the game or from strategy. Now, from watching the game being played he tried to learn two distinct things: the rules of the game and the strategy used in playing it. The game here was Go. The rules are very simple but the strategy very complex; probably more complex than chess. No one in their right mind would claim to know the game of Go if they simply knew the rules—but to play the game they do indeed need to know the rules.

Science is much the same: by observation we try to determine the basic rules and the strategy. In science, the strategy is how the rules manifest themselves in a particular case. We have the rules, for example Newton's laws of motion and gravity. From these, we can deduce the motion of the planets. But just as the rules of Go do not uniquely determine the play, Newton's law do not uniquely determine planetary motion. They describe all possible arrangements of all possible planets. To get to the actual planetary motion we need the initial conditions and a method of getting from the laws to the phenomena. The latter is frequently complicated, and in the case of planetary motion, Newton invented calculus to do the job.

Here we see the two basic aspects of science:

- Reductionism: determining the rules of the game. In physics these rules are frequently called the Lagrangian.
- Emergence: going from the rules to the actual game play. In science, many unexpected phenomena emerge from the basic rules; for example, superconductivity and life.

Albert Einstein (1879 – 1955) was very much a reductionist. He stated: *The grand aim of all science is to cover the greatest number of empirical facts by logical deduction from the smallest number of hypotheses or axioms.* Here is the reductionist's mantra and elusive dream: to reduce everything to a few rules. But it is an elusive dream. We keep pushing to higher energies and get more comprehensive models: Newtonian mechanics, quantum mechanics, quantum field theory, grand-unified models, string theory, whatever is beyond string theory, whatever is beyond whatever is beyond string theory, etc. There is no reason to believe the series will ever end, even if we have it correct up to some level (and string theory may or may not be the correct model, and yes string theory is, at best, just one more model in a string of models).

But even if we knew the rules, we would still need the strategy. Starting with string theory (or even the standard model of particle physics), derive the maximum temperature for superconducting material. For extra points construct the material. Nothing discovered with the Large Hadron Collider (LHC) will help solve this problem.

Henri Poincaré (1854 – 1912), in contrast to Einstein, was not a reductionist[1]:

We seek reality, but what is reality? The physiologists tell us that organisms are formed of cells; the chemists add that cells themselves are formed of atoms. Does this mean that these atoms or these cells constitute reality, or rather the sole reality? The way in which these cells are arranged and from which results the unity of the individual, is not it also a reality much more interesting than that of the isolated elements, and should a naturalist who had never studied the elephant except by means of the microscope think himself sufficiently acquainted with that animal?

Fortunately the blind men who studied the elephant could not use a microscope or they would have had a least one more model to dispute.

Particle physicists tend to think the rules (reductionism) are more fundamental, while condensed matter physicists and chemists think the strategy (emergence) is more fundamental. In both cases fundamental means: *I think what I am doing is way cooler than what you are doing.* And yes, scientists do think what they are doing is cool. If they thought what someone else was doing was cooler they would be doing that. But both reductionism and emergence are necessary for the advancement of science. Anyway I must now go and do some cool, oops! I mean fundamental science.

29. EFFECTIVE FIELD THEORY FOR TURTLES

The story is told (original source unknown) of an elderly woman who attended a talk on Copernicanism. She objected to the speaker, claiming that the he was wrong and that the world was supported on the back of a giant elephant. *And what was the elephant supported by?* asked the speaker. *It stood back of a giant turtle*, replied the lady. Before the speaker could reply she added: *Don't ask what the turtle*

---

[1] Quote from SCIENCE AND HYPOTHESIS.

stands on—I have you there smarty pants, it is turtles all the way down. Now, what has this to do with science? Quite a bit actually.

Albert Einstein (1879 – 1955) once said that the most un-understandable thing about the universe is that it is understandable. But there are two reasons it is understandable. First, the human mind is very well adapted to finding (or creating) patterns and, second, the universe separates itself into different turtles, that is, bite-sized pieces, each characterized by a different scale or size, that can be studied and modeled independently of the other turtles or pieces. Typical scales might be the size of the observable universe, a typical galaxy, the solar system, the earth, people, the atom, the nucleus or the Planck length. One does not have to understand the universe as whole, one can study it one turtle at a time.

For example, I attended a seminar on *ab inito* calculations of nuclear physics where the speaker started with nucleons and the nuclear potential and derived the properties of nuclei without any additional input. *Ab initio* means from the beginning and a member of the audience objected quite strenuously (it had high entertainment value) that this was not *ab inito* because the speaker did not start with quantum chromodynamics (QCD), the assumed underlying model. More pedantically, he should have objected that the speaker should have started with the ultimate theory of everything.

Well, the ultimate theory of everything is not known so where should one start? Obviously, where it is most convenient. In low-energy nuclear physics, this has been a matter of great debate. Historically, the nuclear physicist dealt with nucleons (neutrons and protons) and the interactions between them derived without any reference to an underlying theory (which was unknown at the time). Then along came QCD. The QCD practitioners claimed the nuclear physicists were nincompoops and were wasting their time since they did not start with QCD. This was one of the reasons for the global collapse of nuclear physics in the 1980s. Now it turns out, that by separating the scales using what are called effective field theories, the nuclear physicists were right all along. All the effects of QCD needed for low energy nuclear physics can be accounted for by introducing a few phenomenological parameters to describe the nucleon-nucleon potential (and for purists, many-body forces). Thus there is no need to handle all the complexities of QCD. The QCD practitioners can then calculate the parameters at their leisure. Or not, as the case may be. It really does not matter to the nuclear physicist; all he will ever need is his phenomenological parameters. In the same vein, condensed matter physicists do not have to sit twiddling their thumbs while the particle physicists derive the mass and charge of the electron from theory; they just use phenomenological values. It is the same for other quantities, like the nuclear masses, that condensed matter physicists might need. They use phenomenological values and move on.

This separation into bite-sized pieces happens all the way up and down the set of turtles. Each scale has a different preferred model connected to neighbouring scales and their models by a few parameters. If there are many parameters you have chosen the wrong place to do the separation. In studying ecology, one does not need to know all the chemical interactions. In doing chemistry, one does not need to know all about the quantum mechanical underpinnings. In studying gases, one can determine the volume, pressure and temperature without worrying about the motion of the individual atoms making up the gas. No findings at the LHC will have any effect on biology, chemistry, nuclear physics, or QCD. Except perhaps in developing new experimental or theoretical techniques; they are at vastly different scales. The LHC findings will, however, be crucial for determining the validity of the standard model of particle physics and its extensions.

Deriving the parameters needed at one scale in terms of the smaller scales is reductionism. Sweeping the details of the smaller scales into a few parameters is emergence. There is potentially interesting science at every scale. As always, where one does the division of scales is determined by simplicity and convenience. It is effective field theories (not turtles) all the way down and you can do the separation anywhere you like but if you do it in the wrong place you will be sorry. Cutting turtles in half is messy[1].

---

[1] No turtles were injured in the preparation of this essay.

30. A TALE OF TWO TABLES

Sir Arthur Eddington (1882 – 1944) was one of the leading astrophysicists and publicizers of science in the early to mid-1900s. He measured how much the sun bends light rays and thus helped establish Einstein's general theory of relativity. On the down side, he also proved the fine structure constant was exactly 1/136 and later exactly 1/137; consequently he was referred to as *Sir Arthur Addingone*. His philosophy of science was also suspect, or at least wildly inaccurate.

In his Gifford Lectures of 1927, he talked about two tables. First, the table of everyday experience: it is comparatively permanent, it is coloured, and above all it is substantial. Second, the table of science: it is mostly emptiness with numerous, sparsely-scattered electric charges rushing about with great speed. Eddington's two tables have provided grist for the philosophical mill ever since. Are there really two tables? Susan Stebbing (1885 – 1943) argued that Eddington was mixing everyday language and scientific language in an inadmissible way. But Eddington's crime is much more heinous: he is using the language, appropriate at one scale, to a scale where it is inappropriate. And he is also taking the internals of the models far too seriously. Here is another example (Eddington, 1929)[1]:

> *I am standing on the threshold about to enter a room. It is a complicated business. In the first place I must shove against an atmosphere pressing with a force of fourteen pounds on every square inch of my body. I must make sure of landing on a plank travelling at twenty miles a second round the sun—a fraction of a second too early or too late, the plank would be miles away. I must do this whilst hanging from a round planet, head outward into space, and with a wind of aether blowing at no one knows how many miles a second through every interstice of my body. The plank has no solidity of substance. To step on it is like stepping on a swarm of flies. Shall I not slip through? No, if I make the venture one of the flies hits me and gives a boost up again; I fall again*

---

[1] From THE NATURE OF THE PHYSICAL WORLD (1928), chapter 12.

and am knocked upwards by another fly; and so on. I may hope that the net result will be that I remain about steady; but if unfortunately I should slip through the floor or be boosted too violently up to the ceiling, the occurrence would be, not a violation of the laws of Nature, but a rare coincidence...

Verily, it is easier for a camel to pass through the eye of a needle than for a scientific man to pass through a door. And whether the door be barn door or church door it might be wiser that he should consent to be an ordinary man and walk in rather than wait till all the difficulties involved in a really scientific ingress are resolved.

A complicated business? Only if you use an inappropriate description. A scientific man? Bah! Rather a fool who thinks reductionism is all there is to science. It is striking that twenty years after Einstein's 1905 papers he is still talking about the aether (ether). The atmospheric pressure of the room balances between the front and back so we do not have to *shove* against it. The motion of the earth about the sun is quite irrelevant to the question of entering a room. You can make a poor choice of reference frame (heliocentric rather that geocentric), but do not call it science. The fly analogy is interesting for giving a simple microscopic description of behaviour at the atomic scale but it is a very poor model for describing the large scale. Even as a microscopic description it fails. The electrons in atoms are not moving (technically they are in stationary states) except for thermal motion. And on it goes. Personally, I keep my feet on the ground, and the earth is as solid as it ever was and in that frame the sun also rises. Verily, and in that frame, Joshua could even make it stand still (i.e. it is not logically excluded).

Eddington's is the extreme reductionist's view of the world. If commentating today, he would say all that is real about the elephant can be discovered at the LHC (Large Hadron Collider)—at least until a higher energy accelerator comes along. But there is also emergence: the everyday table is the emergent table and the one I stub my toe on. It is every bit as real as the reductionist's mostly-made-of-emptiness table. Perhaps even more so since the reductionists will always be chasing their elusive table to higher and higher energies, finding yet another new table at each new energy scale: the atomic table (Eddington's), the nuclear table where the nucleus is resolved, the QCD table, the electroweak table, the Planck scale table... and we cannot even speculate intelligently beyond that. If you grant Eddington two tables, you have to grant him many, one at each energy scale; either that or the one at infinite energy which we will never know.

In reality, there is just one table and we know it quite well. However, at each scale we have a preferred, largely self-contained model which we can use to calculate the table's properties. I was about to say *valid model* but I guess a model can be considered valid even if it is too complicated to use in practice. We could, in principle, calculate planetary motion with quantum mechanics, but why bother? For this problem, Newton's laws work as well as they did when he discovered them. Perhaps better, since we now know how to manipulate them more skillfully. Now the mistake Newton—and more especially his disciples—made was to assume that classical mechanics was the ultimate theory of, if not, everything, at least of motion[1]. It may not be the theory of everything, but as a model of slow motion at scales from millimeters to astronomical units it is still valid, as valid as it ever was. Similarly, the everyday table is still a valid concept, as valid as it ever was.

31. IN DEFENSE OF NUCLEAR PHYSICS

Mr. In-Between, Mr. In-Between, Pickin's mighty lean, Mr. In-Between[2].

This song always reminds me of nuclear physics. The scales (i.e. sizes) involved in nuclear physics are too large to be of interest to the reductionists, also known as particle physicists. They say it is just chemistry. The chemists, on the other hand, are not interested because the scales are too small. Nuclear physics, the archetypal in-between

---

[1] Does this remind anyone of the blind men and the elephant?
[2] From a song written by Harlan Howard (1927–2002) and made popular by Burl Ives (1909– 1995).

science, has scales too short to apply directly to everyday life and too long to be at the cutting edge of short-distance physics. In-between science includes atomic physics, low energy nuclear physics, QCD and, if the LHC is successful, electro-weak physics. At the other end of the scale, we have the solar system and galactic science which have too a short a scale to be of interest to the cosmologists who are doing science at scales the size of the visible universe.

So, why do in-between science? Let's take low-energy nuclear physics, the physics done at rare isotope facilities like TRIUMF's Isotope Separator and Accelerator (ISAC) facility, as an example. The nucleus is an intriguing object. It is built of neutrons and protons which are themselves emergent objects, that is, objects that are not present explicitly in the underlying QCD model. They emerge from solving that model. It is somewhat like building on sand, which can be productive and interesting, as in the case of sand castles. Actually, things are not so bad. We now have a very good understanding of the relation between low-energy nuclear physics and QCD.

The nucleus is self-bound: the forces between the components hold it together. This allows all kinds of behaviour: it rotates, vibrates, has single-particle excitations, and pairing. It slices, it dices... well let's not get carried away, this isn't a TV commercial. Disentangling the various types of excitation can be fun—just get any of my experimental colleagues going on the topic. There are real intellectual challenges in sorting it all out. Great progress has been made but we are not at the end of the trail yet.

We also know a lot about nuclear power (no, not in reactors, but in the stars). Stars are powered by gravitationally confined nuclear fusion. No need to build tokamaks[1]—the universe has been powered by nuclear fusion from the beginning. To understand how the universe evolves through time, it is necessary to understand this energy source. And it's not just ordinary stars, but novae and super novae are powered by nuclear energy as well. We are composed of the remnants of stars, remnants blown into space by novae and super novae explo-

---

[1] Devices for producing power from nuclear fusion.

sions. *We are star dust. Billion year old carbon.*[1] To understand all this, is to understand nuclear physics. Explosive, short lived, and dynamic processes in the heavens depend on the properties of short-lived nuclear isotopes. Coming back down to our planet, the need for studying these isotopes and their associated reactions is fulfilled by facilities like ISAC which make and study short-lived isotopes.

Even more down to earth, is nuclear medicine. Medical imaging, using short-lived nuclear isotopes, explores questions such as, *What causes Parkinsonism?* and *Can we catch Alzheimer's disease at an early stage and cure it?* Radiation has been used to cure cancer for a long time now and more progress is being made. In diagnosis and treatment, nuclear medicine is now mainstream. Cyclotrons, once the hallmark of elite physics departments, are now almost a necessity at research hospitals. The pure research in nuclear physics had led to benefits beyond our wildest dreams.

And finally nuclear bombs; destruction beyond our wildest dreams. I would guess that in the USA, the right to keep and bear nuclear arms is covered by the second amendment. In any event, as with any science, nuclear physics can cure or kill. Fire keeps us warm, yet wood smoke is carcinogenic. What we need, always and everywhere, is reality-based thinking and responsible people.

To conclude, in-between science is driven by the same impulse that drives all science: a longing to know and a hope to help. Science at any scale is cool (or is that fundamental?).

I work like a dog with no recreation and they call me Mr In-between
Mr In-between, Mr In-between, makes a fellow mean, Mr In-between[2]

---

[1] From *Woodstock* by Joni Mitchell (b. 1943).
[2] From the earlier referenced song.

32. THE TROUBLE WITH PARTICLE PHYSICS[1]

Ya got trouble, my friend, right here,
I say, trouble right here in River City.[2]

What is the current trouble with particle physics? That's an easy one: a paucity of new experimental results that challenge the status quo. In contrast, in the past twenty years, cosmology has surged ahead, fueled by the new results from COBE[3], WMAP[4], Hubble, and other novel devices. Yet that field may now also be reaching the point of diminishing returns. Without new experimental results any field stagnates. But before addressing this in more detail let's look at some other suggested problems with particle physics.

One of the criticisms of particle physics is the large size of the collaborations. Well that is just the nature of beast. To probe short distances we need large machines. They are expensive and require large collaborations to build, operate and maintain. Being a successful member of a large collaboration requires, in part, a different skill set from that for tabletop science. It relies much more on social skills and no one is required to be a jack-of-all-trades as different members of the collaboration can specialize in different areas. While the skills required may be different they are still as useful to society. The World Wide Web grew out the need for particle physics to collaborate widely. While particle physics still has the largest collaborations, other fields are also moving in that direction. The collaborations needed to build and launch satellite observatories are also large. Even nuclear physics is moving towards large, long time span facilities. While still not in the same league as the ATLAS detector at the LHC, the TIGRESS detector at ISAC (TRIUMF) took seven years to build.

---

[1] Not be confused with Hitchcock's 1955 black comedy: THE TROUBLE WITH HARRY. However, depending on how it turns out, it may indeed be a black comedy.
[2] From the 1963 movie *The Music Man*.
[3] Cosmic Background Explorer.
[4] Wilkinson Microwave Anisotropy Probe.

Other problems were suggested in Lee Smolin's (b. 1955) book THE TROUBLE WITH PHYSICS. (It should have been called THE TROUBLE WITH PARTICLE PHYSICS since it only dealt with that rather small—important but small—part of the totality of physics.) One of his points was that there is too much herd mentality in the field with too many people working on, for example, string theory. To some extent this is a valid objection. Science works best when a variety of different approaches are explored. However, science is self-correcting and trying to impose diversity from the outside is doomed to failure. When there are too many people in one area they sooner or later realize this and some move on. People moving on is the only sure sign that there are too many people in a field. Indeed, this is starting to happen in string theory and will probably turn into a stampede when (hopefully not if) the Large Hadron Collider (LHC) finds surprising new results. He also suggested that particle physics needs more theorists thinking deep thoughts. In my humble opinion, that is the last thing we need—more navel-gazing theoretical particle physicists and this from a long time navel-gazing theorist. What we need are more experimental results so the theorists have something more interesting to gaze at.

Why the shortage of data? Two reasons: The first is not really a shortage of data but a shortage of challenging data. The standard model of particle physics is just too damn (am I allowed to say that?) successful. The detectors at the LHC are starting to churn out data but to date nothing earth shaking. Essentially all tests of the standard model have failed to find anything new, at least at a convincing level. The one possible exception to this is neutrino physics with the underground detector systems. Whether this is an exception depends on how you define the standard model. Independent of that, the neutrino mass and mixing measurements have added excitement to the field with a number of new results, for example the neutrino mixing angle, θ_{13}, from the Tokai to Kamioka (T2K) experiment and Double Chooz.

The second for the shortage of data is the size and time scale of particle-physics projects. For example, the LHC has taken more than 15 years from conception until it produced its first interesting results. T2K has taken a shorter time but it is still many years. The long time scales mean that the exciting new results tend to happen infrequently and the large size also precludes doing things in parallel. This is worrying as it makes independent replication difficult. We have only one large hadron collider. A second was planned but cancelled due to the cost.

A new accelerator, the International Linear Collider (ILC), has been planned and worked on for some time. In 2000, I was assured that by 2006 the construction would have started. *The best-laid schemes o' mice an' men. Gang aft agley*[1]. It has not happened. When it will happen is anyone's guess. Funding a large accelerator project in the current financial situation is going to be tricky.

So LHC, we are relying on you—no pressure or anything. If the LHC finds just the Higgs and nothing unexpected, particle physics will be in tough shape; the dark comedy referred to in the footnote. We have the standard model, which is widely believed to be incomplete, and without unexpected results we have no clue how to go beyond that model—maybe the universe really is fine tuned to many decimal places. Theorists are doing their creative best, but are spinning their wheels. What we need is data to reign in their imaginations.

Not finding the Higgs would have been better (except for the public relations disaster) but even then we would need further experimental indications for what went wrong in order to progress. The best result would be herds of unexpected new particles; barring that, finding particles moving faster than the speed of light would do just fine. Then the trouble with particle physics would be over.

33. THE SECOND LAW OF THERMODYNAMICS AND EVOLUTION

There are some things in science that are just so complicated that they cannot be explained to the uninitiated—things like quantum mechanics, the second law of thermodynamics, how a geek thinks, etc. To understand these things, it takes years of sleeping though dull lectures and late nights carous.... Oops, let's start that again. It takes years of

---

[1] To a Mouse by Robert Burns.

sitting in rapt attention at scintillating lectures, late nights studying (I have it right this time) and the secret initiation rites. Don't forget the secret initiation rites. But in this post, I am going to attempt the impossible and explain the second law of thermodynamics in a way that can be understood by the uninitiated. Fools rush in where angels fear to tread and all that. Now the second law is so complicated that there are several alternate but equivalent formulations. One is due to Rudolf Clausius (1822 – 1888). Now, this is very complicated, so take a very deep breath:

- The second law of thermodynamics (Clausius): *If you want your fridge to work you must plug it in.*

See I told you it was complicated. There is an equally complicated version due to Lord Kelvin (1824 – 1907):

- The second law of thermodynamics (Kelvin): *The exhaust from your car motor will be hot.*

Now you may think I am being facetious but I am not (Okay, maybe about the scintillating lectures). What a fridge does is cool things down by taking heat from the inside and depositing it outside, i.e. it takes heat from where it is cooler (the inside) and deposits it where it is hotter (the outside). An exact statement of the second law is that no process can simply do that: transfer heat from where it is colder to where it is hotter and have no other effect. In the case of the refrigerator, the other effect is turning electricity into heat. No fridge can be 100 percent efficient. Hence, you must plug your fridge in. Similarly, the Lord Kelvin statement is that no heat engine (e.g. your car motor, assuming it is not electric) can be 100 percent efficient and simply turn heat (the burning fuel) into mechanical energy (moving the car). Part of the heat energy must be wasted. In this case, the waste heat is in the hot exhaust.

The second law of thermodynamics, as these two examples illustrate, has significant implications for engineering and was derived in that context. It limits what even the best engineers can do and rules out a large class of second-law-violating perpetual motion machines. Not to be confused with first-law-violating perpetual motion machines which violate energy conservation, the first law of thermodynamics.

You may notice that so far, this discussion has nothing to do with order, disorder or their spontaneous creation or destruction that are so frequently associated with the second law (and used to create confusion and disorder). But there is yet another statement of the second law that does involve order and the concept of entropy. Entropy is simply a way of counting the number of microscopic states that correspond to a given macroscopic state. Think of the air in a room: macroscopically it can be described by the temperature, pressure, and volume. Microscopically, it can be described by the location and motion of all the gazillions of particles that make up the air. Many different locations and motions of the air molecules correspond to the same set of temperature, pressure, and volume. Entropy is related to the number of different locations and motions (technically, the natural logarithm of the number) that correspond to the same temperature, pressure and volume.

One can crudely think of entropy as being the information content of a system—the information needed to specify the location and motions of the gazillions of air molecules in the example above. The lower the entropy, the less information content there is. Also note that ordered systems have less information content than disordered ones. Consider the strings: 1111111111111111 and 1907214836589457. The first is highly ordered and has low information content: it is just all ones. The second needs much more information to describe; each digit must be specified individually. It is less ordered and hence higher information content and entropy.

A third equivalent statement of the second law is that the entropy of a closed system (i.e. one that does not interact with the outside world) can never decrease. Now the earth's atmosphere and biosphere are not closed systems. They get energy from the sun and radiate it back out into space as thermal energy. The atmosphere and the biosphere together act like a giant heat engine using the sun as a source of energy and outer space as a sink for the exhaust heat (like the hot gas from the car engine). This heat engine lifts water from the oceans and puts it on mountaintops. It drives all weather systems including hurricanes and blizzards, evolution, and life itself. It is this heat engine and the lack of thermal equilibrium that generates local regions of low entropy like the sheet of ice on the pond I skated on as child or

86

the alligator in the southern bayou. The alligator, like all living things, has a high degree of order, hence low entropy and information content. Waste heat produced by living things is a side effect of maintaining that order. It might seem strange that I say living things have low information content but it would take much more information to describe in detail the location and motion of the atoms in a homogenized (think blender) alligator[1] than in a living one. There are simply many more ways to arrange the atoms in the homogenized version.

The second law of thermodynamics, being obscure when expressed in terms of entropy, is used to justify all kinds of nonsense. For example, evolution is sometimes claimed to be inconsistent with the second law of thermodynamics. But the second law of thermodynamics is not that obscure: it simply says you must plug your fridge in order for it to work. What's obscure about that? Now, what evolution has to do with plugging in refrigerators is beyond me. After all, the earth's biosphere is plugged directly into solar energy, bypassing the need for the electrical grid and cutting out the middleman. I guess I will have to sleep through some more dull lectures to sort this all out. Zzzzz.

34. CAN SCIENCE ANSWER THE *WHY* QUESTION?

The development of science is often portrayed as a conflict between science and religion, between the natural and the supernatural. But it was equally, if not more so, a conflict with Aristotelian concepts: a change from Aristotle's emphasis on *why* to a dominant role for *how*. To become the mainstream, science had to overcome resistance, first and foremost, from the academic establishment and only secondarily from the church. The former, represented by the disciples of Aristotle and the scholastic tradition, was at least as vociferous in condemning Galileo as the latter. Galileo, starting from when he was a student and for most of his career, was in conflict with the natural philosophers. (I decline to call them scientists.) His conflict with the church was mostly towards the end of his career, after he was fifty and more seri-

---

[1] Relax, I am a theorist and have not homogenized any alligators.

ously when he was nearing seventy. The church itself even relied on the opinions of the natural philosophers to justify condemning the idea the earth moved. In the end science and Galileo's successors won out and Aristotle's natural philosophy was vanquished: the stationary earth, the perfect heavens (circular planetary orbits and perfectly spherical planets), nature abhorring a vacuum, the prime mover and so on. For most of these it is so long and good riddance. So why do philosophers still spend so much time studying a person like Aristotle who got so much wrong? I really don't know.

However, Aristotle did have a few good ideas whose loss is unfortunate. The baby was thrown out with the bath water, so to speak. One such concept, although much abused, is the classification of causes given by Aristotle. The four types of causes he identified are the formal, material, efficient and final causes. He believed that these four causes were necessary and sufficient to explain any phenomena. The formal cause is the plan, the material cause is what it is made of, the efficient cause is the HOW, and the final cause is the WHY. If you think in terms of building a house the formal cause is the blueprint, the material cause is what it is built of (the wood, brick, glass, etc.), the efficient causes are the carpenters and their tools (are hammers obsolete?) and the final cause is the purpose for which the house was built.

Aristotle and his medieval followers emphasized the final cause and pure thought. Science became established only by breaking away from the final cause and the tyranny of WHY. The shift from concentrating on pure thought and the final cause (why) to concentrating on observations and efficient causes (how) was the driving factor in the development of science. Science has now so completely swept Aristotle aside that, at the present time, only the efficient cause is considered a cause in the *cause and effect* sense.

However, in dealing with human activities all four of these types of causes are useful. For example consider TRIUMF where I work. The formal cause is the five-year plan given in a brilliantly written (Okay. I helped write it and they pay my salary so what else could I say) 800-page book that lays out the program for the five years (2010 – 2015) and beyond. The material cause is what TRIUMF is built of (many tons of concrete shielding among other things). The efficient cause is the people and machines that make TRIUMF work. The final cause is TRIUMF's purpose as given in the mission and vision statements. A similar analysis can be done for any organization. The usefulness of the final cause concept is shown by it being resurrected in good management practice under the heading of mission and/or vision statements.

Now, when we go from human activity to animal activity, we lose the formal cause. Consider a bird building a nest. The material cause is what the nest is built of, the efficient cause is the bird itself and the final cause is to provide a safe place to raise its young. But the formal cause does not exist. It is doubtful the bird has a blueprint for the nest; rather the nest is built as the result of efficient causes—the reflexive actions of the bird. No bird ever wrote an 800-page book outlining how to build a nest. Just as well, or the avian dinosaurs (otherwise known as birds) would have gone extinct along with the non-avian ones. A similar analysis exists for simpler organisms. A recent study of yeast showed why (in the sense of the final cause) yeast cells clump together: to increase the efficiency of extracting nutrients from the surroundings. Thus in dealing with human, animal or even yeast activities, science can and does answer the why or final cause question. In the case of the yeast the efficient cause would be the method the yeast cells used to do the bonding, and the material cause, the substances used for the bonding.

When we go from animate to inanimate we lose, in addition to the formal cause, the final cause. Aristotle explained the falling of objects in terms of a final cause: the objects wanted to be at their natural place at the center of the universe, which Aristotle thought was the center of the earth. He thought the reason they sped up as they fell was they became jubilant at approaching their natural place (I am not making that up). Newton, in contrast, proposed an efficient cause: gravity. There was no goal, i.e. final cause, just an efficient cause. A river does not flow with the aim of reaching the sea but just goes where gravity pulls. Similarly with evolution by natural selection, it has no aim but just goes where natural selection pulls. This freaks out those people who insist on formal and final causes. With much ingenuity, they have tried to rectify the situation by proposing formal and final causes: intelligent design and theistic evolution respectively. Intelligent design posits that at least some of the structures found in living organisms are the result of intelligent design by an outside agent and not the result of natural selection while theistic evolution posits that evolution was controlled by God to produce *Homo Sapiens*. Neither has been found to increase the ability of models to make accurate predictions; hence they have no place in science. It is this lack of utility, not the role of a supernatural agent that leads to their rejection as science.

How does the final cause, or at least the illusion of a final cause, arise in living things? It is an emergent property generated by the feedback loop in evolution. Natural selection favors those organisms that respond to stimuli in a manner that looks like purpose: moving towards food sources and away from danger; those that don't leave fewer descendants.

To summarize: for the activities of living things, science can and does answer the why question and assigns a final cause. However, for non-living things science has not found the final cause concept to be useful and has eliminated it based on parsimony. Aristotle, his followers and disciples made the mistake of anthropomorphizing nature and assigning to it causes that are only appropriate to humans or, at best, living things.

35. THE ROLE OF FAITH IN SCIENCE

Back in ancient history, when I was a graduate student, we did not have a computer on every desk. We prepared decks of computer cards and trotted them down to the computer center and waited for the printed output; no getting upset if you did not have one-second response on your monitor. Anyway, I was calculating the same quantity in two different ways. One way involved a complicated calculation solving the Schrodinger equation for many different states and doing an obscure averaging. The other was a much simpler calculation using what is known as semi-classical approximations (to find out more, read my thesis). Relating the two involved a lot of math—calculus, differential equations, Laplace transforms, and various other techniques named after august, dead people. As I sat there looking at the numbers coming out the same, I thought: *Egad, you know, math really does work.* But, is it correct to say I had faith in the validity of mathematics?

Similarly, is my belief in the utility of Newton's laws of motions a matter of faith? But, you say, in this case, my belief or faith is misplaced because the laws are not absolutely correct. So? Where they work, they work extraordinarily well: planetary motion, cars, books, baseballs (except when I am trying to catch them), chalk thrown by an irate teacher (it missed), etc. I do not worry about books starting to move by themselves. Is my belief in the continuing validity of other well-established models and techniques of science a matter of faith or something else? What about evolution, global warming, renormalization techniques, quantum mechanics, and the second law of thermodynamics? Are they all or any of them matters of faith?

The answer to the above question depends on what one means by faith. But if the above are examples of faith it is a rather trite use of the word. Indeed, it is a stretch to claim that any of these are examples of faith at all—certainly not in the same sense that faith is used in religious circles: *Now faith is the assurance of things hoped for, the conviction of things not seen* (Hebrews 11:1). To a large extent, faith, in this latter sense, is absent from science. The rules of engagement are well laid out and there is little need for *a conviction of things not seen.*

But, faith does come into science in two ways: one fundamental to science and the other optional and probably spurious. The spurious one is when science is taken beyond its legitimate bounds and claims are made about the ultimate nature of reality: claims about materialism, naturalism and realism. Here, indeed, we have *a conviction of things not seen.* These are not really a part of science but rather metaphysics and a matter of faith as discussed in a previous essay, LET THE MYSTERY BE. But what about things like atoms, electrons and quarks: things that are not seen in the normal sense of the word but are inferred? They are internal parts of the models science builds and taking them to be definite parts of reality is an act of faith. Like the ether they may go poof at some point in the future. Following Henri

Poincaré (1854 – 1912), Hans Vaihinger (1852 – 1933), and Willard V. O. Quine (1908 – 2000), I rather take their existence as a matter of convention and convenience. Are they *really* there? Who knows. If the math works out the same does it really matter?

At one point faith does play a key role in science. It could be called the fundamental axiom of science or science's Nicene Creed: *Patterns observed in the past enable us to predict what will happen in the future.* The seriousness of the problem was originally pointed out by David Hume (1711 –1776) in his critique of scientific induction. His claim was that scientific induction does not exist. There is no logical reason for tomorrow to be the same as today. Hume had no answer to the problem other than to ignore it. Immanuel Kant (1724 – 1804), responding to Hume, tried to solve the problem and failed.

The fundamental axiom of science lies behind all of science and provides the foundation on which the scientific method rests. We build models based on past observations to predict future observations. This only works if the fundamental axiom is true. The scientific method is just the practical application of this idea. In terms of Aristotle's four types of causes, the scientific method is the formal cause, the scientists the efficient cause, and final cause (the why) is to build models that will correctly predict the future based on the past. The ability to achieve the final cause rests entirely on the fundamental axiom. The formal cause, the scientific method, follows from trying to put the final cause into action. Since the constructs of science are abstract there is no material cause.

The fundamental axiom is a sophisticated version of the *Mount Saint Helens fallacy*. This was named after a person who refused to leave Mount Saint Helens because he did not believe it would blow up. It had not blown up in living memory so why would it blow up now? I am not sure his body was ever found. Today does not have to be like yesterday. But in science we assume the rules will be the same or at least change in predictable ways; not as naïvely as the poor guy on Mount Saint Helens but the assumption is the same: the sun will rise tomorrow (in Vancouver in the winter that is, indeed, *a thing not seen,* due to the clouds). Do the laws of physics tomorrow have to be the same as today? Will mathematics be different tomorrow? Maybe, just maybe, when I look at the two answers on my computer screen[1] tomorrow they will be different. It is possible, but I have faith.

36. THE SIREN CALL OF LOGICAL POSITIVISM

For every problem, there is a simple solution: neat, plausible and wrong.

The philosophers such as Rudolf Carnap (1891 – 1970) and the Vienna Circle considered logical positivism *the received view* of the scientific method. In the early- to mid-twentieth century, it dominated the philosophy of science discussions but is now widely viewed as seriously flawed—or as A. J. Ayer (1910 – 1989), a former advocate, put it: *I suppose the most important [defect]...was that nearly all of it was false.* Pity. But it was good while it lasted. So, what is logical positivism? It is sometimes defined by the statement: *Only verifiable statements have meaning*—note *verifiable* not *falsifiable*. The doctrine included opposition to all metaphysics, especially ontology and synthetic *a priori* propositions. Metaphysics is rejected not as wrong but as having no meaning.

Logical positivism is very nice idea: we work only with observations and what can be deduced directly from them. No need for theories, models or metaphysics. I can hear the cheering now, especially from my experimental colleagues. Logical positivism arose partially in response to the revolutions in physics in the early twentieth century. Quantum mechanics and relativity completely upended the metaphysics and philosophy built around classical mechanics, so the logical positivist wanted to eliminate the metaphysics to prevent this from happening again; a very laudable goal.

So what went wrong? As Ayer noted, almost everything. First, metaphysics tends to be like accents—something only the other person has. The very claim that metaphysics is not needed is itself a metaphysical claim. Second, observations are not simple. As demonstrated by opti-

---

[1] Note computer screen, not computer printout; things have changed.

cal illusions, what we see is not necessarily what is there. The perceptual apparatus does a lot of processing before the results are presented to the conscious mind. The model of the universe presented to the conscious mind probably has more uncontrolled assumptions than any accepted scientific model. But that is what the logical positivists took as the gospel truth. In addition there is Thomas Kuhn's (1922 – 1996) claim that observations are model dependent. While that claim is disputable, it is clear that the interpretation of observations depends on the model, the paradigm or if you prefer the metaphysics; something beyond the observations themselves.

Third as Sir Karl Popper (1902 – 1994) argued, in general, scientific models cannot be verified only falsified. Willard Quine (1908 – 2000) went farther and argued that any statement, in isolation, could be taken as either true or false depending on the background assumptions. Quine's work contributed significantly to the decline and fall of logical positivism. Thus, *only verifiable statements have meaning* would exclude not only all of science but most statements from having meaning. Indeed, it would exclude even that statement itself since the statement: *only verifiable statements have meaning* cannot be verified.

Logical positivism: neat, plausible and wrong. Well can anything be salvaged? Perhaps a little. Consider the statement: In science, only models that can be empirically tested are worth discussing. Not to be overly broad, I restrict the statement to science. The criteria in mathematics are rather different and I do not wish to make a general statement about knowledge, at least not here. Second, I have replaced *statement* with *model* since by the Duhem-Quine thesis individual statements cannot be tested since one can make almost any statement true by varying the supporting assumptions. In the end it is global models that are tested. Science is observationally based, so the adjective empirical. I use tested to avoid complaints about the validity of verification versus falsification. Tested is neutral in that regard. Finally, *meaningful* has been replaced *by worth discussing*. To see why consider the composition of the sun. In the late nineteenth century, it was regarded as something that would never be known. At that point the statement THE SUN IS COMPOSED MAINLY OF HYDROGEN would have been considered meaningless by the logical positivists and cer-

94

tainly, at that time, discussion of the issue would have been futile. But with the discovery of spectroscopic lines, models for the composition of the sun became very testable and the composition of sun is now considered to be well understood. It went from not worth discussing to being well understood but the composition of the sun did not change. I would consider the statement THE SUN IS COMPOSED MAINLY OF HYDROGEN to be meaningful even before it could be tested; meaningful but not worth discussing.

My restatement above does, however, eliminate a lot of nonsense; like the omphalos hypothesis, the flying spaghetti monster, and a lot of metaphysics, from discussion. But its implications are more wide ranging. During my chequered career as a scientist, I have seen many pointless discussions of things that could not be tested: d-state of the deuteron, off-shell properties, nuclear spectroscopic factors and various other technical quantities that appear in the equations used by physicists. There was much heat but little light. It is important to keep track of which aspects of the models we produce are constrained by observation and which are not. Follow the logical positivists, not the yellow brick road, and keep careful track of what can actually be determined by measurements. What is behind the curtain is only interesting if the curtain can be pulled aside.

To conclude: Don't waste your time discussing what can't be empirically tested. That is all that's left of logical positivism once the chaff has been blown away. And good advice it is—except for mathematicians. Either that or I have been lured to the rocks by the siren call of logical positivism and have another statement that is neat, plausible and wrong!

37. IS SCIENCE MERELY FICTION?

Hans Vaihinger (1852 – 1933) was a German philosopher who introduced the idea of AS IF into philosophy. His book, DIE PHILOSOPHIE DES ALS OB (THE PHILOSOPHY OF 'AS IF'), was published in 1911, but written more than thirty years earlier. He seems to have survived the publish or perish paradigm for thirty years.

In his book, Vaihinger argued that we can never know the true underlying reality of the world but only construct systems which we assume match the underlying reality. We proceed AS IF they were true. A prime example is Newtonian mechanics. We know that the underlying assumptions are false—the fixed Euclidean geometry for example—but proceed AS IF they were true and use them to do calculations. The standard model of particle physics also falls into this category. We know that at some level it is false but we use it anyway since it is useful. Vaihinger himself used the example of electrons and protons as things not directly observed but assumed to exist. They are, in short, useful fictions. Willard Quine (1908 – 2000), forty years later, used the same term.

Vaihinger's approach is a good response to Ernst Mach's (1838 – 1916) refusal to believe in atoms because they could not be seen. In the end, Mach lost that fight but not without casualties. His positivism had a negative effect on physics and was a contributing factor in Ludwig Boltzmann's (1844 – 1906) suicide. The philosophy of AS IF is the antithesis of positivism, which holds closely to observation and rejects things like atoms which cannot be directly seen. Even as late as the early twentieth century, some respectable physics journals insisted that atoms be referred to as mathematical fictions. Vaihinger would say to proceed AS IF they were true and not worry about their actual existence. Indeed, calling them mathematical fictions is not far from the philosophy of AS IF.

The ideas of Vaihinger had precursors. Vaihinger drew on Jeremy Bentham's (1748 – 1832) work: THEORY OF FICTIONS. Bentham was the founder of modern utilitarianism and a major influence on John Stuart Mill (1806 – 1873) among others. AS IF is very much a form of utilitarianism: If a concept is useful, use it.

The idea of AS IF was further developed in what is known as fictionalism. According to fictionalism, statements that appear to be descriptions of the world should be understood as cases of MAKE BELIEVE, or pretending to treat something as literally true (a USEFUL FICTION or AS IF). Possible worlds or concepts, regardless of whether they really exist or not, may be usefully discussed. In the extreme case, science is only a useful discussion of fictions; i.e. science is fiction.

The core problem goes back at least to Plato (424/423 BCE – 348/347 BCE) with the parable of the cave (from THE REPUBLIC). There, he talks about prisoners who are chained in a cave and can only see the wall of the cave. A fire behind them casts shadows on the wall and the prisoners perceive these shadows as reality since this is all they know. Plato then argues that philosophers are like a prisoner who is freed from the cave and comes to understand that the shadows on the wall are not reality at all. Unfortunately, Plato (and many philosophers after him) then goes off in the wrong direction. They take ideas in the mind (Plato's ideals) as the true reality. Instead of studying reality, they study the ideals which are reflections of a reflection. While there is more to idealism than this, it is the chasing after a mirage or, rather, the image reflected in a mirage.

Science takes the other tack and says we may only be studying reflections on a wall or a mirage but let us do the best job we can of studying those reflections. What we see is indeed, at best, a pale reflection of reality. The colours we perceive are as much a property of our eyes as of any underlying reality. Even the number of dimensions we perceive may be wrong. String theory seems to have settled on eleven as the correct number of dimensions but that is still in doubt. Thus, science can be thought of as AS IF or fictionalism.

But that is far too pessimistic, even for a cynic like me. The correct metaphor for science is the model. What we build in science are not fictions but models. Like fictions and AS IF, these are not reality and should never be mistaken for such, but models are much more than fictions. They capture a definite aspect of reality and portray how the universe functions. So while we scientists may be studying reflections on a wall, let us do so with the confidence that we are learning real but limited knowledge of how the universe works.

38. IS THERE A PLACE FOR REALISM IN SCIENCE?

In the philosophy of science, realism is used in two related ways. The first way is that the interior constructs of a model refer to something that actually exists in nature, for example does the quantum mechanical wave function correspond to a physical entity. The second way is that properties of a system exist even when they are not being measured; the ball is in the box even when no one can see it (unless it is a relative of Schrodinger's cat). The two concepts are related since one can think of the ball's presence or absence as part of one's model for how balls (or cats) behave.

Despite our (and even a young child's) belief in the continued existence of the ball and that cats are either alive or dead, there are reasons for doubting realism. The three main ones are the history of physics, the role of canonical (unitary) transformations in classical (quantum) mechanics, and Bell's inequality. The second and third of these may seem rather obtuse, but bear with me.

Let's start with the first, the history of physics. Here, we follow in the footsteps of Thomas Kuhn (1922 – 1996). He was probably the first philosopher of science to actually look at the history of science to understand how science works. One of his conclusions was that the interior constructs of models (*paradigms* in his terminology) do not correspond (*refer* in the philosophic jargon) to anything in reality. It is easy to see why. One can think of a sequence of models in the history of physics. Here we consider the Ptolemaic system, Newtonian mechanics, quantum mechanics, relativistic field theory (a combination of quantum mechanics and relativity) and finally quantum gravity. The Ptolemaic system ruled for half a millennium, from the second to seventeenth centuries. By any standard, the Ptolemaic model was a successful scientific model since it made correct predictions for the location of the planets in the night sky. Eventually, however, Newton's dynamical model caused its demise. At the Ptolemaic model's core were the concepts of geocentrism and uniform circular motion. People believed these two aspects of the model corresponded to reality. But Newton changed all that. Uniform circular motion and geocentrism were out and instantaneous gravitation attraction was in. Central to the Newtonian system was the fixed Euclidean space time geometry and particle trajectories. The first of these was rendered obsolete by relativity and the second by quantum mechanics; at least the idea of fixed number of particles survived—until quantum field theory. And if string theory is correct, all those models have the number of dimensions wrong. The internal aspects of well accepted and successful models disappear when new models replace the old. There are other examples. In the history of physics, the caloric theory of heat was successful at one time but caloric vanished when the kinetic theory of heat took over. And on it goes. What is regarded as central to our understanding of how the world works goes poof when new models replace old.

On to the second reason for doubting realism—the role of transformations: canonical and unitary. In both classical and quantum mechanics there are mathematical transformations that change the internals of the calculations[1] but leave not only the observables but also the structure of the calculations invariant. For example, in classical mechanics we can use a canonical transformation to change coordinates without changing the physics. We can express the location of an object using the earth as a reference point or the sun. Now this is quite fun; the choice of coordinates is quite arbitrary. So you want a geocentric system (like Galileo's opponents), no problem. We write the equation of motion in that frame and everyone is happy. But you say the Earth really does go around the sun. That is equivalent to the statement: planetary motion is more simply described in the heliocentric frame. We can go on from there and use coordinates as weird as you like to match religious or personal preconceptions. In quantum mechanics the transformations have even more surprising implications. You would think something like the correlations between particles would be observable and a part of reality. But that is not the case. The correlations depend on how you do your calculation and can be changed at will with unitary transformations. It is thus with a lot of things that you might think are parts of reality but are, as we say, model dependent.

Finally we come to Bell's inequality as the third reason to doubt realism. The idea here goes back to what is known as the Einstein-Podolsky-Rosen paradox (published in 1935). By looking at the correlations of coupled particles Albert Einstein (1879 – 1955), Boris

---

[1] For the relation between the two types of transformations see: N.L. Balazs and B.K. Jennings, Unitary Transformations, Weyl's Association and the Role of Canonical Transformations, PHYSICA, 121A (1983) 576–586.

Podolsky (1896 – 1966), and Nathan Rosen (1905 – 1995) claimed that quantum mechanics is incomplete. John Bell (1928 – 1990), building on their work, developed a set of inequalities that allowed a precise experimental test of the Einstein-Podolsky-Rosen claim. The experimental test has been performed and the quantum mechanical prediction confirmed. This ruled out all local realistic models; that is, local models where a system has definite values of a property even when that property has not been measured. This is using realism in the second sense defined above. There are claims, not universally accepted, that extensions of Bell's inequalities rule out all realist models, local or non-local.

So where does this leave us? Pretty much with the concept of realism in science in tatters. (It can, of course, always be recovered by the judicious use of Duhem-Quine thesis, for example by redefining realism[1]). The internals of models change in unpredictable ways when science advances. Even within a given model, the internals can be changed with mathematical tricks and for some definitions of realism, experiment has largely ruled it out. Thus we are left with our models that describe aspects of reality but should never be mistaken for reality itself. Immanuel Kant (1724 – 1804), the great German philosopher, would not be surprised[2].

But what about the ultimate nature of reality? There is no theorem that says reality, itself, must be simple as the scientific models assumes; so scientific models imply very little about the ultimate nature of reality. I guess we will have to leave that discussion to the philosophers and theologians. More power to them.

39. OPERATIONALISM AND OPERATIONAL DEFINITIONS

Physicists frequently stray into the field of philosophy; notable examples include Thomas Kuhn (1922 – 1996) and Henri Poincaré (1854 – 1912). This is perhaps because physicists frequently work in

---

[1] See, for example, Structural Realism.
[2] He made the distinction between the thing in itself and observations of it.

areas far removed from everyday experiences where, in order to understand and be successful in communicating their ideas, underlying assumptions must be dealt with explicitly. Although less well known today than Kuhn and Poincaré, Percy Bridgman (1882 – 1961) also falls into this category. In physics, he is noted for his work on high-pressure physics, winning the Nobel Prize in 1946. In philosophy, he is credited with coining the term *operational definition* and promoting the idea of operationalism. These ideas are laid out in the 1927 book: THE LOGIC OF MODERN PHYSICS. If nothing else, it shows the folly of using *MODERN* in book titles. Nonetheless, it is an interesting little book and, in its time, quite influential.

In his book, Bridgman introduces several interesting ideas, for example, that when one explores new areas in science, one should not be surprised that the supporting concepts have to change. Hence we should not be surprised when classical concepts fail in the relativistic or quantum domains. This illustrates why interpretations of quantum mechanics, explaining it in terms of classical concepts, are poorly motivated. A related idea is that an explanation is the description of a given phenomenon in terms of familiar concepts. Of course, with this definition, what qualifies as a valid explanation depends on what the explainee is familiar with. But what happens when there are no such familiar concepts available? Bridgman suggests that the solution is to introduce new concepts and become familiar with them. Seems reasonable to me. Thus quantum mechanics can be explained in terms of the, familiar to me, concept of the wave function; no need for many worlds and the like.

While it is natural to think of high speed (relativity) or small size (quantum mechanics) as new areas of science, Bridgman includes increased precision as well. He talks about the penumbra of uncertainty that surrounds all measurements and that is penetrated by increasing the precision of the measurements. Thus the idea of the distinct high-energy and precision frontiers, commonly discussed in modern particle physics planning exercises, goes back at least to 1927.

Bridgman was also a phenomenologist to the core. He did not believe that *a priori* knowledge could constrain what could happen; in his words: *Experience is determined only by experience.* C.I. Lewis

(1983 – 1964), in his 1929 book Mind and the World Order, agrees. The similar ideas, in books of about the same time, indicate the concerns of that age.

Despite these interesting asides, the main idea in The Logic of Modern Physics is that concepts are defined by how they are measured; that is by the measurement operation, hence the term operationalism. So why was he interested in operational definitions? It was to avoid the problem in classical mechanics where concepts like distance and time were taken for granted. It then came as a shock when the concepts proved to be rather complex when special relativity was invented. To avoid such shocks in the future, Bridgman proposed the idea of operational definitions. For example, to measure length you go down to the local Canadian Tire® store (in the USA it would be Walmart®), buy a tape measure and use it to measure length. Thus the concept of length is defined by Canadian Tire®, oops, I mean by a tape measure. What if I measure length by surveying techniques that make use of tranquilization? Bridgman claimed that that is a distinctly different concept and is covered by the same term only for convenience. Here at TRIUMF, distance and location are also measured using laser tracking. This is again different from the original concept of length. Things get even more complicated when we talk about the distance to stars which again use a different operation. Bridgman suggests that length loses it meaning at lengths less than the size the electron because such lengths cannot be measured. Today we would say they can be measured but length in that case is simply a parameter in a mathematical formula describing the scattering of particles. Hence we do not have one concept of length or distance but many, although they are the same numerically in regions where the techniques overlap.

Bridgman then goes on to consider various other concepts and how they might be defined operationally. He seems to have been very much influenced by Albert Einstein (1879 – 1955) and Einstein's discussion of the synchronization of clocks (which actually goes back to Poincaré). The possible operational definitions of velocity are particularly interesting. In contradistinction to the definition given by Einstein based on clocks synchronized and distances measured in a fixed inertial frame, Bridgman suggests that the velocity of a car could also be defined by counting mileposts that the car passes to determine distance and using the clock on the car dashboard to measure time. This velocity can become infinite and would be useful to a person going to a distant solar system who is interested in how many of his years it takes to get there. For most purposes Einstein's definition is more convenient and hence it is the one in textbooks though other definitions remain possible.

And on it goes. In some cases his definitions seemed quite contrived. Nevertheless, three groups of people picked up on the idea of operational definitions. One group was the logical positivists. They tried to avoid theory and were pleased when a physicist gave definitions directly in terms of observables. The second group was the psychologists, who wanted a more secure foundation for their subject. The third group was in quality control and business management where Walter Shewhart (1891 – 1967) and Edwards Deming (1900 – 1993) adopted the idea. Of the three, it is only the last where the influence has continued until today. Deming devotes a chapter to it in a 1982 book: OUT OF CRISIS.

However the concept, as the end all and be all of meaning, had its problems. Like logical positivism, it missed the idea that the meaning is in the model. While we may have different ways to measure length there is common idea behind them all. We can consider this common idea to be an abstraction from the different operationally defined concepts or we can take the operational definitions as approximations to the abstract idea. One could say that operationally there is no difference between these two approaches.

Ultimately, despite their shortcomings, operational definitions are useful. They tie concepts tightly to observations where they are less likely to be dislodged by future discoveries or new models. They also help eliminate fuzzy thinking. For example, the scientific method provides a clean unambiguous operational definition of knowledge in contrast to the usual definition as justified true belief. A lot of the concepts that do not have operational definitions are, in general, poorly defined. Who knows, I might even take the concept of scientific realism seriously if someone gave me an operational definition of it.

Here I sit with my feet up thinking deep thoughts and some fool tells me I have to publish if I want to get paid. Curse you, Robert Boyle (1627 – 1692)! Don't they know that publishing takes time away from thinking deep thoughts? After all, thinking deep thoughts is the sole reason for theoretical physicists to exist. Ah well, I suppose, I must. But things were different before Robert Boyle. Alchemists were very careful about who could learn their secrets. A lot of the information was passed down orally to apprentices or written in code. After all, if you had learned the magic incantation for turning lead into gold, you did not want your competitors butting in and driving up the price of lead. But that all changed with Robert Boyle. He started the trend of publishing his results so others could build on what he had done. Also perhaps because he could not, himself, understand what his assistant, Robert Hooke (1635 – 1703), had done and thought that others perhaps could if he made the results available. Thus he published and started a trend.

Since the time of Robert Boyle, publishing has become the standard by which scientists are judged. One needs at least one publication for a Ph. D, 15 to 20 for a permanent job (in my specialty), and one very good one to get a Nobel Prize. Unfortunately, publishing does not directly correlate with how much you get paid. Now, my father cut down trees for a living and was paid for each ton of trees trucked to the pulp mill. Perhaps it could be similar with scientists: pay them by the ton of paper consumed rather than produced. At the end of the year, weigh up the paper used to publish their work and pay accordingly. A good journal would then be one that had a large circulation and used a lot of dead trees. Of course, then you might get a bunch of really prolific writers, but a lack of deep thinkers. And never mind electronic publication—that throws a whole new element in. Guess we'll have to throw the paid-by-the-ton scheme out the window.

The world of electronic publishing leads to an interesting digression: What is a publication? Everyone agrees that words printed on dead trees and circulated form a publication. But what about words that never appear on dead trees? With even Newsweek becoming an electronic only publication, I guess that electronic publications must be considered legitimate. Going further, what about preprint servers like arXiv (a widely-used preprint server that contributed significantly to the end of paper preprints)? It seems to me that arXiv largely replaces the need for the traditional journals. I always took the point of view that I would put my papers on arXiv so other scientists would read them and then submit it to a regular journal so I could put it on my CV as a peer-reviewed publication. I also used that electronic archive as my main source of information on what was going on in my field. The archival, printed journals I rarely looked at. If we had a rating and peer-review system for papers on the electronic archives we could safely do away with the traditional journals. However, my boss does like to brag about the number of laboratory papers that make the cover of NATURE.

But back to the main point, publishing is important. The first reason is that while publishing a lot of papers does not necessarily indicate that one is making a major contribution, no papers probably does indicate that one is sleeping rather than thinking deep thoughts. Thus, papers published should be considered the first indication of scientific productivity—and a baseline for your supervisor to keep paying your or not. The second reason (and the one that Boyle initiated when he didn't understand his assistant's work) is that peer review, in the broad sense, plays a major role in error control. It is one's peers that will ultimately decide if one's thoughts are deep or shallow, on track or not. The only way one's peers can critique one's work is if it is published and made available. The third reason is that science progresses by building on what has gone before, and to this, we must thank Boyle. It is the published papers and the much-maligned archival journals that keep the record of what has been learned. While much that is published can safely be forgotten, the gems, like Einstein's papers, are also there.

So if you wish to flourish as a scientist and not perish, it is best to publish—but only good papers, as to not bog down the archives or kill too many trees. As for me, I wonder if I can count these blogs as publications on my CV. That would give me an additional eighty publications. Probably not. Anyway when I retire in a few years, I will have a CV burning party[1] so it really does not matter.

41. GRANTS AND THE SCIENTIFIC METHOD

Many years ago, I served on a committee responsible for recommending funding levels for research grants. After the awards were announced, a colleague commented that all we did was count the number of publications and award grants in proportion to that number. So, I checked and did a scatter plot. Boy, did they scatter. The correlation between the grant size and the number of publications was not that strong. I then tried citations; again a large scatter. Well, perhaps the results really were random—nah, that could not happen; I was on the committee after all.

I did not do a multivariable analysis, but there were no simple correlations between what might be called quantitative indicators and the size of the research grant. This supports the conclusions of the EX-PERT PANEL ON SCIENCE PERFORMANCE AND RESEARCH FUNDING: *Mapping research funding allocation directly to quantitative indicators is far too simplistic, and is not a realistic strategy*[2]. Trying to do that is making the mistake of the logical positivists who wanted to attach significance directly to the measurements. As I have argued in previous essays, the meaning is always in the model and logical positivism leads to a dead end.

In deciding funding levels, the situation is too complicated for the use of a simple algorithm. Consider the number of publications. There are different types of publications: letters, regular journal articles, review articles, conference contributions, etc. Publications are of different lengths. Should one count pages rather than publications? Or is one letter worth two regular journal papers; letters being shorter and con-

---

[1] It has been suggested the CV should be shredded not burnt to avoid carbon loading.
[2] Quote from: INFORMING RESEARCH CHOICES: INDICATORS AND JUDGMENT. Expert Panel on Science Performance and Research Funding. Council of Canadian Academies (2012).

sidered by some to be more important than regular articles. But, in reality, one wants to see a mix of the different types of publications. A review article might indicate standing in the field but one also wants to see original papers. Is a paper in a *prestigious* journal worth more than one in a more mundane journal? What is a *prestigious* journal anyway? There is also the question of multi-author papers. One gets suspicious if all the papers are with more senior or well-known authors but all single author papers is also a warning sign. Generally co-authoring papers with junior collaborators is a good thing. In some fields, all papers include all members of the collaboration so the number of coauthors carries very little information. The order of authors on a publication may or may not be important. And on it goes. Expert judgment is, as always, required to sort out what it all means.

Citations are an even bigger can of worms. Even in a field as small as sub-atomic theoretical physics there are distinct variations in the pattern of citations among the subfields: string theory, particle phenomenology or nuclear physics. For example, the lifetime for citations in particle phenomenology is significantly less than in nuclear physics. Then there is the question of self-citations, citations to one's own work or, more subtle, to close collaborators. And what about review articles? Is a citation to a review article as important as one to an article on original research? Review articles frequently collect more references. My most cited paper is a review article. A person can, with a bit of effort, sort this all out. Setting up an algorithm would be damn near impossible. A person could even, gasp, read some of the papers and form an independent opinion of their validity. But that could introduce biases. Hence, numbers are important but they must be interpreted. This leads to the conclusion: *Quantitative indicators should be used to inform rather than replace expert judgment in the context of science assessment for research funding allocation.*[1]

The other problem with simple algorithms is the feedback loop. With a simple algorithm, researchers naturally change their behaviour to maximize their grants. For example, if we judge on the number of

---

[1] *Ibid*

publications, people split papers up, publish weak papers, or publish what is basically the same thing several times. I have done that myself. None of these improve the quality of the work being done. Expert judgment can generally spot these a mile away. After all, the experts have used these tricks themselves.

More generally, there is the problem of trying to reduce everything to questions that have nice quantitative answers. *Far better an approximate answer to the right question, which is often vague, than an exact answer to the wrong questions, which can always be made precise[1].* There seems to be this argument that since science normally uses quantitative methods, administration should follow suit so it can have the success of science. It is like the medieval argument that since most successful farmers had three cows; the way to make farmers successful was to give them all three cows. But the wrong question can never give the right answer. It is far better to ask the right question and then work on getting a meaningful answer. What we want to do at a science laboratory, or for funding science generally, is to advance our understanding of how the universe works to the maximum extent possible and use the findings for the benefit of society. The real question is how do we do this? That is neither an easy question to answer nor one that can be easily quantified. Not being quantifiable does not make it a meaningless question. There are various metrics an informed observer can use to make intelligent judgments. But it is very important that administrators avoid the siren call of logical positivism and not try to attach meaning directly to a few simple measurements.

42. THE ORIGINS OF SCIENCE: THE AUTHORIZED VERSION

The true origins of science are lost in the mists of time. Possibly it started when some Australopithecus observed that a stick with a knot at the end was more effective in warding off a rival for its[2] mate than

---

[1] Tukey, J. W. (1962). The Future of Data Analysis. Annals of Mathematical Statistics 33(1), 1-67.

[2] Note, gender neutral pronoun.

one without a knot. Since then the use of the scientific method has occasionally intruded into mainstream life but until the seventeenth century was always beaten back into the ground by philosophers, theocrats, and the proponents of common sense: Of course the earth is flat[1] and no, stones do not fall from the sky. But in the seventeenth century science *took* and began its path to mainstream acceptance. To be definite, I would take the date for the emergence of science to be that night in 1609 when Galileo first pointed his telescope to the heavens. Two questions then present themselves: 1) Why was it so late in the advance of civilization that science arose and 2) Why did it arise when and where it did? The first question was addressed in the first essay and the second will be addressed here.

The date chosen for the beginning of science is rather arbitrary since science did not spring full blown out of nothing. There were precursors and aftershocks but the early seventeenth century is as good a starting point as any. And it was just not in astronomy. In 1600, William Gilbert, (1544 – 1603) published his work on magnetism, DE MAGNETE, MAGNETICISQUE CORPORIBUS, ET DE MAGNO MAGNETE TELLURE, and in 1628 William Harvey (1578 – 1657) released his work on blood circulation, DE MOTU CORDIS. Back in astronomy, Johannes Kepler (1571 – 1630) published his first book on elliptic planetary orbits in 1609 (1609 was indeed a propitious year) and a multivolume astronomy textbook about ten years later. So back to the basic question: why so much activity then and there?

There are a number of different reasons. The first is a slow accumulation of ideas that suddenly reached a critical point and took off. Even Galileo Galilei's father, Vincenzo Galilei, played a role. He helped put music theory on an empirical and mathematical basis, and influenced his son towards applied mathematics. Inventions also played a role; for Galileo to point his telescope at the heavens, the telescope had first to be invented. Besides the telescope, the invention of the printing press around 1440 by Johannes Gutenberg[2] played a key role. It greatly increased the ease with which new ideas could propagate. It played a key role in the Protestant Reformation and allowed Nicolaus

---

[1] In British Columbia and Switzerland, it is crinkly rather than flat.
[2] Although the Koreans may have invented it earlier.

Copernicus's (1473 – 1543) ideas of a heliocentric planetary system to spread throughout Europe. It also played a key role in disseminating Galileo's ideas.

But there is more than that. In the thirteenth century, Western Europe began to rediscover the ancient learning of the Greeks, especially Aristotle. This came by way of the Arab world which added original contributions (*e.g.* Arabic numerals and chemistry in the form of alchemy) to the store of knowledge and also collected information from other sources, for example, India (*e.g.* zero). Building on that foundation, Western Europe built an academic tradition at universities and monasteries. This mostly consisted of scholasticism and the worship of Aristotle but it did set the stage for intellectual debate and the pursuit of knowledge as an end in itself. In the end, science destroyed the scholasticism and the worship of Aristotle that had laid part of the foundation for its success.

The rediscovery of Greek learning in the thirteenth century had a surprising side effect. The Greeks were long on rational thought, but regarded the things of the world as changing and unpredictable, probably due to their belief in capricious Gods and fickle Fates. Christian Europe believed in a supreme, omnipotent God. This led at least one part of the church to regard science, the study of how the world worked, as sacrilegious since it seemed to imply a limit on what God could do. But combine the Greek ideal of rationalism with the idea of an omnipotent being and suddenly things change. The very concept of perfect was taken to imply rationality. Hence, the perfect God must be rational and create a rational and ordered universe; namely one in which it made sense to look for orderly laws. Indeed, in nineteenth century England, it was a common belief that God ruled through orderly natural laws. And of course, it was the scientist's role to discover these laws. Religion played another role. The Protestant Reformation was a shift from the authority of a man, namely the Pope, to the written word of the Bible. Science was also a shift from the authority of a person, Aristotle, to the unwritten word, namely the universe. The people of the time talked of God's word and God's work and considered both worthy of study; study without the need for a human intermediary.

The Protestant Reformation also destroyed a source of central authority—the Catholic Church. This, coupled with political fragmentation (especially in Germany) led to more change. There was no longer any central authority to suppress new ideas, yet enough rule of law to allow fairly rapid communication (again thanks in part due to the printing press). For example, Galileo's works were published by the Jewish publishing house of Elsevier in protestant Holland while he was under house arrest in Italy by the Catholic Church.

As noted in the first paragraph, science has had from the beginning three main opponents (using anachronistic terms): the academic left, the religious right, and common sense. For Galileo, the academic left was represented by the natural philosophers, the religious right by the Catholic Church that the philosophers sicced on him, and common sense by those who *knew* heavier objects fell faster than light ones. At various times, different ones of these have been predominant: editorials attacked the idea of rocks falling from the sky (meteorites) or rockets working in space were there was no air to react against. In the 1960s, the main opponents were the academic left with the idea that scientific laws were mostly, if not entirely, cultural and postmodernism remains an opponent of science. But today, the main opposition to science comes from the religious right with evolution being the main fall guy. But same three protagonists—the academic left, the religious right and common sense—have remained and will probably remain into the indefinite future as the main opponents of science. *As it was in the beginning, is now, and ever shall be. World without end. Amen*[1].

43. THE ORIGINS OF SCIENCE: THE UNAUTHORIZED VERSION

In any genealogy there are always things one wants to hide; the misfit relative, the children born on the wrong side of the sheet, or the relative Aunt Martha just does not like. As a genealogist it takes a lot of effort to find these things out. Genealogies tend to be sanitized: the

---

[1] From the DOXOLOGY known as *Gloria Parti*.

illegitimate grandchild becomes a *legitimate* child, the misfit relative somehow missed being included, and Uncle Ben aged 15 years during the 10 years between censuses and died at age one hundred although he was only eighty[1] according to the earliest records. The roots of science, like all family histories, have undergone a similar sanitization process. In the previous essay, I gave the sanitized version. In this essay, I give the unauthorized version. Aunt Martha would not be happy.

The Authorized Version has the origins of science tied closely to the Greek philosophical tradition with science arising within and from that tradition. While that has some truth to it, it is not the whole truth. It was two millennia from Aristotle to the rise of science and there have been many rationalizations for this delay, many starring Christianity as the culprit. However, Christianity did not gain political strength for a few centuries after Aristotle's death. So, if the cause was Christianity, it must have had a miraculous non-causal effect.

A lot happened between Aristotle and the rise of science: the rise and fall of the Roman Empire, the rise of Christianity and Islam, the Renaissance, the reformation, and the printing press. A lot of what could be called knowledge was developed but did not provide major gains in philosophy. The Romans were for the most part engineers not philosophers, but to do engineering takes real knowledge. Let us pass on to the Arabs. They are generally considered as a mere repository for Greek knowledge which was then passed on to the West largely intact but with some added commentary. I suspect that is not correct, as I will argue shortly.

There are two contributions to the development of science that are frequently downplayed: astrology and alchemy. These are the ancestors that science wants to hide. We all know the story of the Ptolemy and Copernicus, but the motivation for the development of astronomy was astrological and religious. From the ancient Babylonians to the present day, people have tried to divine the future by studying the stars. It is no accident that astronomy was one of the first sciences. It

---

[1] My genealogy of the Musquodoboit Valley has all these and more (http://www.rootsweb.ancestry.com/~canns/mus.pdf).

had *practical* applications: astrology (Kepler was a noted astrologer) and the calculation of religious holidays. One of the reasons Copernicus's book was not banned was because the church found it useful for calculating the date of Easter. The motion of the planets is also sufficiently complicated that they could not be predicted trivially, yet they were sufficiently simple to be amenable to treatment by the mathematics of the day. Hence, it became the gold standard of science. Essentially we dropped the motivation but kept the calculations.

Alchemy, the other problem ancestor, is even more interesting. The Arabs, those people who are considered to have produced nothing new, had within their ranks Jābir ibn Hayyān (721 – 815), chemist and alchemist, astronomer and astrologer, engineer, geographer, philosopher, and physicist, pharmacist and physician—in general, an all-round genius. He, along with Robert Boyle (1627 – 1691), is regarded as the founder of modern chemistry, but note how far in advance Jābir ibn Hayyān was, about 900 years. Certainly he took alchemy beyond the occult to the practical. Although the alchemists never succeeded in turning lead into gold, they did produce a lot of useful metallurgy and chemistry. It is indeed possible that along with his chemical pursuits, Jābir ibn Hayyān forged the foundation of science by going down to the laboratory and seeing how things actually worked. Following in Hayyān's footsteps, Ibn al-Haytham (c. 965 – c. 1040) is possibly the father of modern optics, ophthalmology, experimental physics and scientific methodology, and the first theoretical physicist. Not a bad resume It is no surprise that the first two people to introduce something like science into Western Europe, Frederick II (1194 – 1250) and Roger Bacon (c. 1214 – 1294), were both very familiar with Arab scholarship and presumably with Jābir ibn Hayyān's and Ibn al-Haytham's work. In addition, both Isaac Newton (1642 – 1727) and Robert Boyle (1627 – 1691) were alchemists. A major role for alchemy in the development of science cannot be creditably denied.

It is perhaps wrong to think of astrology and alchemy as separate. To turn cabbage into sauerkraut, you need to know the phase of the moon[1] and the same probably holds true for turning lead into gold. Hence, most alchemists were also astrologers. But alchemy and astrology have always had a dark side of occult, showmanship, and outright fraud. A typical, perhaps apocryphal, example would be Dr. Johann Georg Faust (c. 1480 – c. 1540). He was killed around 1540 when his laboratory exploded. Or his laboratory exploded when the devil came to collect his soul. This person is presumably the origin of the Dr. Faust legend of the man who sold his soul to the devil for knowledge. It interesting that this legend arose in the late sixteenth century just as science was beginning to rise from obscurity. The general population's suspicion of learning, there from the beginning, has perhaps never really gone away.

The philosophers and theologians, the beautiful people, had their jobs in the monasteries and universities but science owes more to the people who sold their soul, or at least their health, to the devil for knowledge. These were the people who went down to the laboratories and did the dirty work to see how the world actually works.

44. SHUT UP AND CALCULATE

Andreas Osiander (1498 – 1552) was a Lutheran theologian who is best remembered today for his preface to Nicolaus Copernicus's (1473 – 1543) book on heliocentric astronomy: DE REVOLUTIONIBUS ORBIUM COELESTIUM. The preface, originally anonymous, suggested that the model described in the book was not necessarily true, or even probable, but was useful for computational purposes. Whatever motivated the Lutheran Osiander, it was certainly not keeping the Pope and the Catholic Church happy. It might have been theological, or it could have been the more general idea that one should not mix mathematics with reality. Johannes Kepler (1571 – 1630), whose work provided a foundation for Isaac Newton's theory of gravity, took Copernicus's idea as physical and was criticized by no less than his mentor, Michael Maestlin (1550 – 1631) for mixing astronomy and physics. This was all part of a more general debate about whether or not the mathematical descriptions of the heavens should be consid-

---

[1] Or so my uncle claimed.

ered merely mathematical tricks or if physics should be attached to them.

Osiander's approach has been adopted by many others down through the history of science. Sir Isaac Newton—the great Sir Isaac Newton himself—did not like action at a distance and when asked about gravity said, *Hypotheses non fingo.* This can be roughly paraphrased into English as: *shut up and calculate.* He was following Osiander's example. It was not until Einstein's general theory of relativity that one could do better. Even then, one could take a shut up and calculate approach to the curved space-time of general relativity.

Although atoms were widely used in chemistry, they were not accepted by many in the physics community until after Einstein's work on Brownian motion in 1905. Ernst Mach (1838 – 1916) opposed them because they could not be seen. Even in the early years of the twentieth century Mach and his followers insisted that papers discussing atoms, published in some leading European physics journals, have an Osiander-like introduction. And so it continues: in his first paper on quarks, Murray Gell-Mann (b. 1929) introduced quarks as a mathematical trick. If Alfred Wegener (1880 – 1930) had used that approach to continental drift it might not have taken fifty years for it to be accepted.

We see a trend: ideas that are considered heretical or at least unorthodox—heliocentrism, action at a distance, atoms, and quarks—are introduced first as mathematical tricks. Later, once people become used to the idea, they take on a physical reality, at least in people's minds. In one case, the trend went the other way. Maxwell's equations describe electromagnetic phenomena very well. They are also wave equations. Now, physicists had encountered wave equations before and every time, there was a medium for the waves. Not being content to shut up and calculate, they invented the ether as the medium for the waves. Lord Kelvin (1824 – 1907) even proposed that particles of matter were vortices in the ether. High school textbooks defined physics in terms of vibrations in the ether. And then it all went *poof* when Einstein published the special theory of relativity. Sometimes, it is best to just shut up and calculate.

Of course, the expression *shut up and calculate* is applied most notably to quantum mechanics. In much the same vein as with the ether, physicists invented the Omphalos ... oops, I mean the many-worlds interpretation of quantum mechanics to try to give the mathematics a physical interpretation. At least Philip Gosse (1810 –1888), with the Omphalos hypothesis, only had one universe pop into existence without any direct evidence of the pop. The proponents of the many-worlds interpretation have many universes popping into existence every time a measurement is made. Unless someone comes up with a subtle knife[1] so one can travel from one of these universes to another, they should not be taken any more seriously than the ether.

The *shut up and calculate* approach to science is known as instrumentalism[2]—the idea that the models of science are only instruments that allow one to describe and predict observations. The other extreme is realism—the idea that the entities in the scientific models refer to something that is present in reality. Considering the history of science, the role of simplicity, and the implications of quantum mechanics[3] (a topic for another essay), realism—at least in its naïve form—is not tenable. Every time there is a paradigm change or major advance in science, what changes is the nature of reality given in the models. For example, with the advent of special relativity, the fixed space-time that was a part of reality in classical mechanics vanished. But with an instrumentalist's view, all that changes with a paradigm change is the range of validity of the previous models. Classical mechanics is still valid as an instrument to predict, for example, planetary motion. Indeed, even the caloric model of heat is still a good instrument to understand many properties of thermodynamics and the efficiency of heat engines. Instrumentalism thus circumvents one of the frequent charges against science: namely that we claim to know how the universe works and then discover that we were wrong. This is only true if you take realism seriously and apply it to the internals of models.

---

[1] *The Subtle Knife*, the second novel in the *His Dark Materials* trilogy, was written by the English novelist Philip Pullman.
[2] A term introduced by the pragmatic philosopher John Dewey (1859 – 1952).
[3] In particular Bell's inequalities.

The model building approach to science advocated in these essays is an intermediate between the extremes of instrumentalism and realism. The models are judged by their usefulness as instruments to describe past observations and make predictions for new ones; hence the tie-in to instrumentalism. The models are not reality any more than a model boat is, but they capture some not-completely-determined aspect of reality. Thus, the models are more than mere instruments, but less than complete reality. In any event, one never goes wrong by shutting up and calculating.

45. THE INTERPRETATION OF QUANTUM MECHANICS

When I first started dabbling in the dark side and told people I was working on the philosophy of science, the most common response from my colleagues was: *oh the foundations of quantum mechanics?* Actually not. For the most part, I find the foundations of quantum mechanics rather boring. Perhaps that is because my view of science has a strong instrumentalist tinge, but the foundations of quantum mechanics have always seemed to me to be trying to fit a quantum reality into a classical framework; the proverbial triangular peg in an hexagonal hole. Take wave-particle duality for example. Waves and particles are classical idealizations. The classical point particle does not exist, even within the context of classical mechanics. It should come as no surprise that when the classical framework breaks down, the concepts from classical mechanics are no longer valid. What quantum mechanics is telling us is only that the classical concepts of waves and particles are no longer valid. Interesting, but nothing to get excited about.

The problem with the uncertainty principle is similar. This principle states that we cannot simultaneously measure the position and motion of a particle. Now, classically, the state of a particle is given by its location and motion (i.e. it's momentum). Quantum mechanically, the state is given by the wave function or, if you prefer, by a distribution in the location-motion space[1]. Now the problem is not that the location and motion cannot be measured simultaneously but that the particle does not simultaneous have a well-defined position and motion since its state is given by a distribution. This causes realists, at least classical realists, to have fits. In quantum mechanics, the position is only known when it is directly measured, i.e. properties of the system only exist when they are being looked at. This is a distinctly anti-realist point of view. Again, this is trying to force a classical framework on a quantum system. If anything is real in quantum systems, it is the wave functions, not individual observables. But see below.

Quantum mechanics is definitely weird; it goes against our common sense, our intuition. The main problem is that, while classical mechanics is deterministic, quantum mechanics is probabilistic. To see why this is a problem, consider the classical-probability problem of rolling a dice. I roll a fair dice. The chance of it being 2 is 1/6; similarly for any value from 1 to 6. Now once I look at the dice the probability distribution collapses. Let's say, I see a 2. The probability is now 1 that the value is 2 and zero for the other values. But for Alice who has not seen me check, the probabilities are still all 1/6. I now tell her that the number is even. This collapses her probability distribution so that it is 1/3 for 2,4,6 and zero for 1,3,5. Now for Bob, who did not hear me telling Alice, the probabilities are still 1/6 for each of the numbers. Two important points arise from this. First, classical probabilities change discontinuously when measurements are made and, second, classical probabilities depend not just on the system but on the observer, i.e. probabilities are observer dependent.

We should expect the same quantum mechanically. We should expect measurements to discontinuously change the probability distribution and the probability distribution to be observer dependent. The first is certainly true. Quantum mechanical measurements cause the wave function to collapse and consequently the probability distribution[2] also collapses.

---

[1] Technically, the phase space.
[2] The probability is the absolute value of the wave function squared.

118

The second point is not commonly realized or accepted, but it should be. The idea that the wave function is a property of the quantum system plus observer, not the quantum system in isolation, is not new. Indeed, it is a variant of the original Copenhagen interpretation of quantum mechanics. But frequently, it is denied. When this is done, one is usually forced to the conclusion that the mind or consciousness plays a large and mysterious role in the measurement process. Making the wave function or the state description observer dependent avoids this problem. The wave function is then just the information the observer has about the quantum system. As Niels Bohr (1885 – 1962), one of the founders of quantum mechanics, said: *It is wrong to think that the task of physics is to find out how nature is. Physics concerns what we can say about nature.*

Let us consider the wave function collapse in more detail. Consider an entanglement experiment. The idea is to have a system emit two particles such that if we know the properties of one, the properties of the other are also known. One of the two emitted particles is measured by Bob and the other by Alice.[1] Now, Alice is lazy so she has her particle transported to her home laboratory. She also knows that once Bob has done his measurement, she does not have to measure her particle but only has to call Bob to get the answer. Bob is also lazy, but he does go to the lab and, if he feels like it, does the measurement and faithfully records it in his log book. One day when Alice calls, she gets no answer. It turns out Bob has died between the time he would have made the measurement and when he would have recorded it in his lab book. Now Alice is very upset. Not that Bob has died—she never liked him anyway—but that she does not know if the momentous event of the wave function collapse has happened or not. Her particle has not arrived at her home yet, but there is no experiment she can do on it to determine if the wave function has collapsed or not. The universe may have split into many worlds but she can never know! Of course, if the wave function is a property of the observer-quantum system, there is no problem. The information Bob

---

[1] By convention it has to be Bob and Alice. I believe this is a quantum effect.

had on the wave function was lost when Bob died and Alice's wave function is as it always was. Nothing to see here, move along.

So what is the interpretation of quantum mechanics? An important part seems to be that wave function is the information the observer has on the quantum system, and is not a property of the quantum system alone. If you do not like that, well there is always instrumentalism,[1] *i.e.* shut up and calculate.

46. REALITY AND THE INTERPRETATIONS OF QUANTUM MECHANICS

If there were only one credible interpretation of quantum mechanics, then we could take it as a reliable representation of reality. But when there are many, it destroys the credibility of all of them. The plethora of interpretations of quantum mechanics lends credence to the thesis that science tells us nothing about the ultimate nature of reality.

Quantum mechanics, in its essence, is a mathematical formalism with an algorithm for how to connect the formalism to observation or experiments. When relativistic extensions are included, it provides the framework for all of physics[2] and the underlying foundation for chemistry. For macroscopic objects (things like footballs, deflated or not), it reduces to classical mechanics through some rather subtle mathematics, but it still provides the underlying framework even there. Despite its empirical success, quantum mechanics is not consistent with our commonsense ideas of how the world should work. It is inherently probabilistic despite the best efforts of motivated and ingenious people to make it deterministic. It has superposition and interference of the different states of particles, something not seen for macroscopic objects. If it is weird to us, just imagine how weird it must have seemed to the people who invented it. They were trained in the classical system until it was second nature and then nature itself

---

[1] Instrumentalism has no problem with quantum mechanics or, indeed, any other scientific model.
[2] Although quantum gravity is still a big problem.

said, *Fooled you, that is not how things are.* Some, like Albert Einstein (1879 – 1955), resisted it to their dying days.

The developers of quantum mechanics, in their efforts to come to grips with quantum weirdness, invented interpretations that tried to understand quantum mechanics in a way that was less disturbing to common sense and their classical training. In my classes in quantum mechanics, there were hand waving discussions of the Copenhagen interpretation, but I could never see what they added to mathematical formalism. I am not convinced my lecturers could either, although the term *Copenhagen interpretation* was uttered with much reverence. Then I heard a lecture by Sir Rudolf Peierls[1] (1907 – 1995) claiming that the conscious mind caused the collapse of the wave function. That was an interesting take on quantum mechanics, which was also espoused by John von Neumann (1903 – 1957) and Eugene Wigner (1902 – 1995) for part of their careers.

So does consciousness play a crucial role in quantum mechanics? Not according to Hugh Everett III (1930 – 1982) who invented the many-worlds interpretation. In this interpretation, the wave function corresponds to physical reality, and each time a measurement is made the universe splits into many different universes corresponding to each possible outcome of the quantum measurement process. Physicists are nothing if not imaginative. This interpretation also offers the promise of eternal life. The claim is that in all the possible quantum universes there must be one in which you will live forever. Eventually that will be the only one you will be aware of. But as with the Greek legend of Tithonus, there is no promise of eternal youth. The results may not be pretty.

If you do not like either of those interpretations of quantum mechanics, well have I got an interpretation for you. It goes under the title of the relation interpretation. Here the wave function is simply the information a given observer has about the quantum system and may be different for different observers; nothing mystical here and no multiplicity of worlds. Then there is the theological interpretation. This I

---

[1] A major player in the development of quantum many body theory and nuclear physics.

first heard from Steven Hawking (b. 1942) although I doubt he believed it. In this interpretation, God uses quantum indeterminacy to hide his direct involvement in the unfolding of the universe. He simply manipulates the results of quantum measurements to suit his own goals. Well, He does work in mysterious ways after all.

I will not bore you with all possible interpretations and their permutations. Life is too short for that, but we are still left with the overarching question: which interpretation is the one true interpretation? What is the nature of reality implied by quantum mechanics? Does the universe split into many? Does consciousness play a central role? Is the wave function simply information? Does God hide in quantum indeterminacy?

Experiment cannot sort this out since all the interpretations pretty much agree on the results of experiments (even this is subject to debate), but science has one other criteria: parsimony. We eliminate unnecessary assumptions. When applied to interpretations of quantum mechanics, parsimony seems to favour the relational interpretation. But, in reality, parsimony, carefully applied, favours something else; the instrumentalist approach. That is, don't worry about the interpretations, just shut up and calculate. All the interpretations have additional assumptions not required by observations.

47. NOBODY UNDERSTANDS QUANTUM MECHANICS? NONSENSE!

Despite the old canard about nobody understanding quantum mechanics, physicists do understand it. With all of the interpretations ever conceived for quantum mechanics, this claim may seem a bit of a stretch, but like the proverbial ostrich with its head in the sand, many physicists prefer to claim they do not understand quantum mechanics, rather than just admit that it is what it is and move on.

What is it about quantum mechanics that generates so much controversy and even had Albert Einstein (1879 – 1955) refusing to accept it? There are three points about quantum mechanics that generate controversy. It is probabilistic, it eschews realism, and it is local. Let us look at these three points in more detail.

Quantum mechanics is probabilistic, not determinist. Consider a radioactive atom. It is impossible, within the confines of quantum mechanics, to predict when an individual atom will decay. There is no measurement or series of measurements that can be made on a given atom to allow me to predict when it will decay. I can calculate the probability of when it will decay or the time it takes half of a sample to decay but not the exact time a given atom will decay. This lack of ability to predict exact outcomes, but only probabilities, permeates all of quantum mechanics. No possible set of measurements on the initial state of a system allows one to predict precisely the result of all possible experiments on that state.

Quantum mechanics eschews realism[1]. This is a corollary of the first point. A quantum mechanical system does not have well-defined values for properties that have not been directly measured. This has been compared to the moon only existing when someone is looking at it. For deterministic systems one can always safely infer back from a measurement what the system was like before the measurement. Hence if I measure a particle's position and motion I can infer not only where it will go but where it has come from. The probabilistic nature of quantum mechanics prevents this backward looking inference. If I measure the spin of an atom, there is no certainty that is had only that value before the measurement. It is this aspect of quantum mechanics that most disturbs people, but quantum mechanics is what it is.

Quantum mechanics is local. To be precise, no action at point A will have an *observable* effect at point B that is instantaneous, or non-causal. Note the word observable. Locality is often denied in an attempt to circumvent Point 2 (realism), but when restricted to what is observable, locality holds. Despite the Pentagon's best efforts, no messages have been sent using quantum non-locality.

---

[1] Realism as defined in the paper by Einstein, Podolsky and Rosen, Physical Review 47 (10): 777–780 (1935).

Realism, at least, is a common aspect of the macroscopic world. Even a baby quickly learns that the ball is behind the box even when he cannot see it. But much about the macroscopic world is not obviously determinist, the weather in Vancouver for example (it is snowing as I write this). Nevertheless, we cling to determinism and realism like a child to his security blanket. It seems to me that determinism or realism, if they exist, would be at least as hard to understand as their lack. There is no theorem that states the universe should be deterministic and not probabilistic or vice versa. Perhaps god, contrary to Einstein's assertion, does indeed like a good game of craps.

So quantum mechanics, at least at the surface level, has features many do not like. What has the response been? They have followed the example set by Philip Gosse (1810 – 1888) with the Omphalos hypothesis[1]. Gosse, being a literal Christian, had trouble with the geological evidence that the world was older than 6,000 years, so he came up with an interpretation of history that the world was created only 6,000 years ago but in such a manner that it appeared much older. This can be called an interpretation of history because it leaves all predictions for observations intact but changes the internal aspects of the model so that they match his preconceived ideas. To some extent, Tycho Brahe (1546 – 1601) used the same technique to keep the earth at the center of the universe. He had the earth fixed and the sun circle the earth and the other planets the sun. With the information available at the time, this was consistent with all observations.

The general technique is to adjust those aspects of the model that are not constrained by observation to make it conform to one's ideas of how the universe should behave. In quantum mechanics these efforts are called interpretations. Hugh Everett (1930 – 1982) proposed many worlds in an attempt to make quantum mechanics deterministic and realistic. But it was only in the unobservable parts of the interpretation that this was achieved and the results of experiments in this world are still unpredictable. Louis de Broglie (1892 – 1987) and later David Bohm (1917 – 1992) introduced pilot waves in an effort to restore realism and determinism. In doing do they gave up locality.

---

[1] See the essay *Science and Simplicity* for more details.

Like Gosse's work, theirs was a nice proof in principle that, with sufficient ingenuity, the universe could be made to conform to almost any preconceived ideas, or at least appear to do so. Reassuring I guess, but like Gosse it was done by introducing non-observable aspects to the model: not just unobserved but in principle unobservable. The observable aspects of the universe, at least as far as quantum mechanics is correct, are as stated in the three points above: probabilistic, nonrealistic and local.

Me, I am not convinced that there is anything to understand about quantum mechanics beyond the rules for its use given in standard quantum mechanics textbooks. However, interpretations of quantum mechanics might, possibly might, suggest different ways to tackle unsolved problems like quantum gravity and they do give one something to discuss after one has had a few beers (or is that a few too many beers).

48. THE ROLE OF THE INDIVIDUAL IN SCIENCE AND RELIGION

Lady Hope (1842 – 1922)[1] in 1915 published a claim that Charles Darwin (1809 – 1882) on his death bed had recanted his views on evolution and God. This story published thirty-three years after Darwin's death was strongly denied by his family but has made the rounds of various creationist publications and websites to this day. Now my question is: Why would anyone care? It may be of interest to historians but nothing Darwin wrote, said, or did has any consequences for evolution today. The theory itself and the evidence supporting it have moved far beyond Darwin. But this story does serve to highlight the different role of individuals in science as compared to religion or even philosophy.

I have always considered it strange that philosophy places such importance[2] on reading the works of long dead people—Aristotle, Descartes, etc. In science, Newton's ideas trumped those of both Aristotle

---

[1] Otherwise known as Elizabeth Reid nee Cotton.
[2] Perhaps it is because, unlike science, philosophy is not accumulative.

and Descartes, yet very few scientists today read Newton's works. His ideas have been taken, clarified, reworked, and simplified. The same thing applies to the scientific writings of other great and long dead scientists. Nothing is gained by going to the older sources. Science advances and the older writings lose their pedagogical value. This is because in science, the ultimate authority is not a person, but observation.

A given person may play an important role but there is always someone else close on his heels. Natural selection was first suggested, not by Darwin, but by Patrick Matthew (1790 – 1874) in 1831 and perhaps by others even earlier. Alfred Russell Wallace's (1823 – 1913) and Darwin's works were presented together to the Linnean Society in July 1858[1]. And so it goes: Henri Poincaré (1854 – 1912) and Hendrik Lorentz (1853 – 1928) were nipping at Einstein's heels when he published his work on special relativity. Someone gets priority, but it is observation that ultimately should be given the credit for new models.

When the ultimate role of observation is forgotten, science stagnates. Take, for example, British physics after Isaac Newton (1642 – 1727). It fell behind the progress on the continent because the British physicists were too enamored of Newton. But the most egregious example is Aristotle (384 BCE – 322 BCE). The adoration of Aristotle delayed the development of knowledge for close to two millennia. Galileo (1564 – 1642) and his critic, Fortunio Liceti (1577 – 1657), disputed about which of them was the better Aristotelian, as if this were the crucial issue. Even today, post-docs all too frequently worry about what the supervisor means rather than thinking for themselves: *But he is a great man, so his remark must be significant*[2]. Actually he puts his pants on one leg at a time like anyone else.

Then there is the related problem of rejecting results due to their origins, or the associated ideology. The most notorious example is the Nazi rejection of non-Aryan science; for example, relativity because

---

[1] The president of the Linnean Society remarked in May 1859 that the year had not been marked by any revolutionary discoveries.
[2] I have heard that very comment.

Einstein was a Jew. One sees a similar thing in politics where ideas are rejected as being socialist, fascist, atheist, Islamic, Christian, or un-American thus avoiding the real issues of the validity of the idea: *Darwinism*[1] *is atheistic hence it must be condemned.* Yeah? And your mother wears army boots.

In science, people are considered great because of the greatness of the models they develop or the experimental results they obtained. In religion, it is the other way around. Religions are considered great based on the greatness of their founder. Jesus Christ is central to Christianity: *and if Christ has not been raised, then our preaching is vain, your faith also is vain* (1 Corinthians 15:14). Islam is based on the idea: *There is no God but Allah and Mohammad is his prophet.* Many other major religions (or philosophies of life) are founded on one person: Moses (Judaism), Buddha (Buddhism), Confucius (Confucianism), Lao Tzu (Taoism), Guru Nanak (Sikhism), Zoroaster (Zoroastrianism), Bahá'u'lláh (Bahá'í Faith) and Joseph Smith (Mormonism). Even at an operational level, certain people have an elevated position and are considered authorities: for example, the Pope in the Catholic Church, or the Grand Ayatollahs in Shi'ite Islam. Because of the basic difference between science and religion, an attack on a founder of a religion is an attack on its core, while an attack on a scientist is an irrelevancy. If Joseph Smith (1805 – 1844) was a fraud, then Mormonism collapses. Yet nothing in evolution depends on Darwin, nor anything in classical mechanics on Newton, hence the irrelevance of Lady Hope's claim. But we can understand the upset of the Islamic community when Mohammad is denigrated: it is an attack on the whole Islamic religious framework which depends on Mohammad's unique role.

The difference in the role of the individual in science and of the role of the individual in religion is due to the different epistemologies of science and religion. In science, everything is public—both the observations and the models built on them. In contradistinction, the inspiration or revelation of religion is inherently private, a point noted by Saint Thomas Aquinas (1225 – 1274). You too can check Ein-

---

[1] Note also the attempt to associate evolution with one person.

127

stein's calculations or Eddington's experiment; you do not have to rely on either Einstein or Eddington. Now it may take years of hard work and a lot of money, but in principle it can be done. But you cannot similarly check the claims of Jesus's divinity, even with years of study, but must take it on faith or as the result of private revelation.

Unlike in science, in religion, old is better than new. If a physical manuscript of St. Paul's writing dating from the first century were discovered, it would have a profound effect on Christianity. But a whole suitcase of newly discover works in Newton's or Darwin's handwriting would have no effect on the progress of science. This is because religion is based on following the teachings of the inspired leader, while science is based on observation.

49. THE ROLE OF AUTHORITY IN SCIENCE AND IN LAW

In the eleventh century, Western Europe rediscovered the teachings of ancient Greece. Two friars played a lead role in this: the Dominican Saint Thomas Aquinas (1225 – 1274) and the Franciscan Roger Bacon (1214/1220 –1292). Aquinas combined the teaching of Aristotle with Christianity. His teachings became the orthodoxy in both Christianity and natural philosophy until the scientific revolution in the fifteenth century. Aquinas took Aristotle as an authority and, in turn, was taken as an authority by those who followed him. To some extent this has continued down to the present day, at least in the Catholic Church. The scientific revolution was, to a large extent, the overturning of Aristotelian philosophy as repackaged by Aquinas.

Bacon took a different track and extracted something different from the study of Aristotle and Arab knowledge. This something different was an early version of the scientific method. He applied mathematics to describe observations and advocated using observation to test models. Bacon described a repeating cycle of observation, hypothesis, experimentation, and the need for independent verification. Bacon was largely ignored and, unlike Aquinas, was not declared a saint. Galileo Galilei (1564 – 1642), if not directly influenced by Bacon, was in many ways following his tradition, both in his use of mathematics and in stressing the importance of observations. The difference between Aquinas and Bacon is the contrast between the appeal to authority and the finding out for oneself. In this contest, the appeal to authority lost rather decisively, but it was a long tough fight. People generally prefer a given answer, even if it is wrong, to the tough process of extracting the correct answer.

In spite of all that, appeal to authority is frequently necessary. The legal system in most democracies, for example, is based on the idea of appeal to authority. The parliament may make the laws but it is the courts that decide on what they mean. Frequently, the courts even have the authority to override laws based on the constitution. This is true in many countries but most famously in the United States of America. In these countries, what the Supreme Court says, is the law. What a law actually means is commonly a matter of interpretation as evidenced by split decisions where one judge holds one opinion and another judge the opposite. Perhaps the interpretations are even arbitrary as they sometimes change over time despite the authority given to precedence. But a decision is required and there is no objective criteria, so the majority on the court rules.

Now, it is worth commenting that laws of nature and laws of man are completely different beasts and it is unfortunate that they are given the same name. The so called laws of nature are descriptive. They describe regularities that have been observed in nature. They have no prescriptive value. In contrast, the laws of man are prescriptive, not descriptive. Certainly, the laws against smoking marijuana are not descriptive in British Columbia, neither were the laws against drinking during US prohibition. The laws describe what the government thinks should happen with prescribed punishments for those who disobey. However, there is no penalty for breaking the law of gravity because, as far as we know, it can't be done. If someone actually did it, it would cease to be a law and there would be a Nobel Prize, not a penalty. Like the laws of man, the laws of God—for example, the Ten Commandments—are prescriptive, not descriptive, with penalties given for breaking them. You can break the law of man and the laws of God, but not the laws of physics.

In science, things are different than in the courts of law. In the latter, we are concerned with the meaning of a law that some group of people have written. This, by its very nature, has a subjective component. In science, we are trying to discover regularities in how the universe operates. In this, we have the two objective criteria: parsimony and the ability to make correct predictions for observations. As pointed out in the previous essay, idolizing a person is a mistake, even if that person is Isaac Newton (1642 – 1727). Appeal to observation trumps appeal to a human authority, but in the short term, even in science, appeal to a human authority is often necessary. Life is too short and the amount to know too large to discover it all for oneself. Thus, one relies on authorities. I consult the literature rather than trying to do experiments myself. We consult other people for expertise that we do not have ourselves. We rely on the collective wisdom of the community as reflected in the published literature. When we require decisions, we must rely on the proximate authorities of peers in a process called peer review. This process is relied on to maintain the collective wisdom and will be discussed in more detail in the next essay. In the meantime, we conclude this essay by paraphrasing William Lyon McKenzie King[1] (1874 – 1950): *Appeal to authority if necessary but not necessarily appeal to authority.*

50. PEER REVIEW: A CORNERSTONE OF SCIENCE

Ah yes, peer review; one of the more misunderstood parts of the scientific method. Peer review is frequently treated as an incantation to separate the wheat from the chaff. What has been peer reviewed is good; what hasn't is bad. But life is never so simple. In the late 1960s, Joseph Weber (1919 – 2000) published two Physical Review Letters were he claimed to have detected gravitational waves. Although there still are a few holdouts who believe he did, the general consensus is that he did not, since his results have not been reproduced. Rather it is generally believed that his results were an experimental artifact. His results were peer reviewed and accepted at a *prestigious* journal but

---

[1] The longest serving Canadian Prime Minister.

that does not guarantee that they are correct. Even the Nobel committee occasionally makes mistakes, most notably giving the award to the discoverer of lobotomies.

Conversely, consider the case of Alfred Wegener (1880 – 1930). In 1912 he proposed the idea of continental drift. To say the least, it was not enthusiastically received. It did not help that Wegener was a meteorologist, not a geologist. This theory was largely rejected by his peers in geology. For example, the University of Chicago geologist Rollin T. Chamberlin said, *If we are to believe in Wegener's hypothesis we must forget everything which has been learned in the past 70 years and start all over again.* In 1926, the American Association of Petroleum Geologists (AAPG) held a special symposium on the hypothesis of *continental drift* and rejected it. After that, the hypothesis was strictly on the fringe until the late 1950s and early 1960s when it finally became accepted.

Thus, we see that peer review cannot *definitively* be relied on to give the final answer. So what use is peer review? The problem is that, as pointed out in previous essays, in science there is no one person who can serve as the ultimate authority; rather, observation can and does. As a school student, the teacher knows more than the student and can be considered the final authority. In university, the professor plays that role, sometimes with gusto. But when it comes to research, frequently it is the researcher himself or herself who is the world expert. So how can research be judged and how do we make decisions about that research? And decisions do have to be made. We cannot publish everything—the useful results would get lost in the noise. We must maintain the collective wisdom that has been laboriously developed. Similarly, decisions have to be made on who gets research grants. Do we use a random number generator? Okay, no snide remarks, I admit that it does occasionally look like we do. As there is no single human to serve as the final authority, we turn to the people who know the most about the topic, namely the peers of the person. If we want a decision related to sheep farming, we consult sheep farmers; if about nuclear physics, we consult nuclear physicists. Peer review is simply the idea that when we have to make a decision, we consult those people most likely to be able to make an informed decision. Is it perfect?

No. Is there a better process? Perhaps, but no one seems to know what it is.

Peer review is also used as a bulwark against bull..., oops, material, that is of questionable validity. The expression, *that has not been peer reviewed,* is used as a euphemism for, *that is complete and utter crap and I am not going to waste my time dealing with it.* In this case it tends to come across as closed minded: *Not peer reviewed? It's nonsense!* Needless to say, cranks take great exception and tend to regard peer review as a new priesthood who stifles innovation. And indeed, as noted above, sometimes peer review does get it wrong. There is always this tension between accepting nonsense and rejecting the next big thing. As the case of continental drift illustrates, it is sometimes only in retrospect, when we have more data, that we can tell what the correct answer is. However, it is better to reject or delay the acceptance of something that has a good chance of being wrong than to have the literature overrun with wrong results (think lobotomies). However, contrary to popular conception, Copernicus and Wegener are the exception, not the rule. That is why Copernicus is still used as the example of the suppression of ideas half a millennium later—there are just not that many good examples. And I might add that both Copernicus and Wegener were initially rejected for good reasons and were accepted once sufficient supporting data came to light. Most people whom the peer review process deems to be cranks, are indeed cranks. Never heard of Immanuel Velikovsky[1] (1895 – 1979)? Well, there is a reason. The few who were right are remembered, but the multitudes who were wrong are, like Velikovsky, mercifully forgotten.

Peer review is one of the cornerstones of science and is an essential part of its error control process. At every level in science we use peers to check for errors. Within well run collaborations, results are reviewed by the peers within the collaboration before submitting for publication. I will get my peers to read my papers before submission.

---

[1] The author of the bestselling and controversial book, Worlds in Collision, (Macmillan. ISBN 1-199-84874-3, 1950). It had an unusual history for planetary motion.

Even the editing of these essays before being put online can be considered peer review. Then there is the formal peer review a paper receives when it is submitted to a journal. In many ways this is the least important peer review because it is after a paper is published that it receives its most vigorous peer review. I can be quite sure there is no fundamental flaw in special relativity, not because Einstein was a genius, not because it was published in a prestigious journal, but because after it was published many very clever people tried very hard to find flaws in it and failed. Any widely read scientific paper will be subject to this thorough scrutiny by the author's peers. That is the reason we can have confidence in the results of science and why secrecy is the enemy of scientific progress. *Given enough eyeballs, all bugs are shallow*[1].

51. ERROR CONTROL IN SCIENCE

Scientists are subject to two contradictory criticisms. The first is that they are too confident of their results, to the point of arrogance. The second is that they are too obsessed about error control—all this nonsense about double blind trials, sterilized test tubes, lab coats and the like. It evidently has not occurred to the critics that the reason scientists are confident of their results is that they have obsessed over error control. Or conversely they obsess over error control so they can be confident of their results.

Now, most people outside science do not realize a scientist's day job is error control. There is this conception of scientists having brilliant ideas, going into immaculate labs where they effortlessly confirm their results to the chagrin of their competitors. That is, of course, when they are not plotting world domination like Pinky and the Brain. But scientists neither spend their time plotting world domination (be with you in a minute Brain) nor doing effortless experiments. But rather they are thinking about what might be wrong; how do I control that error? As for theorists, they must be a part of a wicked and adul-

---

[1] Quotes from THE CATHEDRAL AND THE BAZAAR by Eric Raymond.

terous generation[1] because they are always seeking after a sign—a minus sign that is.

So what do scientists do to control errors? There are very few arrows in their quiver, really only three: care in doing the experiment or calculation, care in doing the experiment or calculation, and care in doing the experiment or calculation. Well actually there are two others: peer review and independent repetition. Let's take the first three first: care, care, and care. As previously noted, scientists are frequently criticized here. Why do double blind studies when we have Aunt Martha's word for it that Fyffe's Patented Moose Juice cured her lumbago? Well actually, testimonials are notoriously unreliable. A book[2] I have, had an example from the early 1900s of testimonials for cures for consumption and then had the dates the person died of consumption. The death was frequently quite close to the date of the testimonial. So no, I will not trust Aunt Martha's testimonial[3]. To quote Robert L. Park (b. 1931)[4]: *The most important discovery of modern medicine is not vaccines or antibiotics, it is the randomized double-blind test, by means of which we know what works and what doesn't.* This has now carried over into subatomic physics where *blind* analyses are common. By blind, I mean that the people doing the analysis cannot tell how close they are to the expected answer (the theoretically predicted answer or the results of a previous experiment) until most of the analysis has been completed. Otherwise, as one of my experimental colleagues said: data points are like sheep, they travel in flocks. Even small biases can influence the results. Blind analysis is just one example of the extremes scientists go to, to ensure that their results are reliable. All this rigmarole that scientists go through is one of the reasons life expectancy increased by about 30 years between 1900 and 2000, perhaps the major reason. The lack of this care is the reason I distrust alternative medicine.

---

[1] Matthew 12:39,16:4.
[2] PSEUDOSCIENCE AND THE PARANORMAL, Terence Hines, Prometheus Books, Buffalo (1988).
[3] My grandfather died of consumption.
[4] A former director of public information at the Washington office of the American Physical Society.

We now move on to the other two aspects of error control: peer review and independent replication of results. Both of these depend on the results being made public. Since these are crucial to error control, results that have not been made available for scrutiny should be treated with suspicion. Peer review has been discussed in the previous essay and is just the idea that new results should be run past the people who are most knowledgeable so they can check for errors.

Replication is, in the end, the most important part of error control. Scientists are human, they make mistakes, they are deluded, and they cheat. It is only through attempted replication that errors, delusions, and outright fraud can be caught. And it is very good at catching them. In the next essay, I will go into the examples but it is a good practice not to trust any exciting new result until it has been independently confirmed. However replication and reproducibility are not simple concepts. I go outdoors and it is nice and sunny, I go out twelve hours later and it is dark and cold. The initial observation is not reproduced. I look up, I see stars. An hour later I go out and the stars are in different places. And the planets, over time, they wander hither, thither and yon. In a very real sense the observations are not reproduced. It is only within the context of model or paradigm that we can understand what reproducible means. The models, either Ptolemaic or Newtonian, tell us where to look for the planets and we can reproducibly check they are where the models say they should be at any given time. Reproducibility is always checking against a model prediction.

Replication is also not just doing the same things over and over again. Then you would make the same mistakes and get the same results over and over again. You do things differently, guided by the model being tested, to see if the effect observed is an artifact of the experimental procedures or real. Is there really a net energy gain or have you just measured a hot spot in your flask. The presence of the hot spot can be reproduced, but put in a stirrer to test the idea of energy gain and, damn, the effect went away. Another beautiful model was slain by an ugly observation. Oh, well, happens all the time.

So science advances as we keep testing our previous results in new and inventive ways. The wrong results fall by the wayside and are forgotten. The correct ones pile up and we progress. To err is human, to control errors—science.

52. MOST EXCITING NEW RESULTS ARE WRONG!

Giovanni Schiaparelli (1835 – 1910) is mainly remembered for his discovery of *canali* on Mars. What a fate, to be remembered only for discovering something that does not exist. I suppose it could be worse; he could be remembered only as the uncle of the fashion designer, Elsa Schiaparelli (1890 – 1973). But he was not alone in seeing canals on Mars. The first recorded instant of the use of the word *canali* was by Angelo Secchi (1818 – 1878) in 1858. The canals were also seen by William Pickering (1818 – 1878) and most famously by Percival Lowell (1855 – 1916). That's right, the very Lowell for whom the Lowell observatories on Mars Hill Road in Arizona are named. Unfortunately, after about 1910, better telescopes failed to find the canals and they faded away. Either that or Marvin the Martian filled them in while fleeing from Bugs Bunny. However, they did provide part of the backdrop for H.G. Wells's WAR OF THE WORLDS. But it is interesting that the canals were observed by more than one person before being shown to be optical illusions.

Another famous illusion (delusion?) was the n-ray discovered by Prof. Blondlot (1844 – 1930) in 1903. These remarkable rays, named after his birth place, Nancy, France, could be refracted by aluminum prisms to show spectral lines. One of their more amazing properties was that they were only observed in France, not England or Germany. About 120 scientists in 300 papers claimed to have observed them (note the infallibility of peer review). But then Prof. Robert Wood (1868 – 1955), at the instigation of the magazine Nature, visited the laboratory. By judiciously and surreptitiously removing and reinserting the aluminum prism, he was able to show that the effect was physiological, not physical. And that was the end of n-rays and also of poor Prof. Blondlot's reputation.

Probably the most infamous example of nonsense masquerading as science is *Homo piltdownensis,* otherwise known as the Piltdown man. This was the English answer to the Neanderthal man and the Cro-Magnon man discovered on the continent. A sculptured elephant bone, found nearby, was even jokingly referred to as a cricket bat. Seems appropriate. While there was some skepticism of the find, the powers that be declared it to be a breakthrough and it was only forty years later that someone had the brilliant idea that it might be a fake. Once the signs of faking were looked for, they were easily found. What we see here is an unholy combination of fraud, delusion, and people latching onto something that confirmed their preconceived ideas.

These examples are not unique. Most exciting new results are wrong[1]: polywater, the 17 kev neutrino, cold fusion, superheavy element 118, pentaquarks, and the science on almost any evening news cast. Cancer has been cured so often it is a surprise that any cancer cells are left. So, why so many exciting wrong results? First, to be exciting means, almost by definition, that the results have a low prior probability of being correct. The discovery of a slug eating my garden plants is not exiting—annoying—but not exciting. It is what we expect to happen. But a triceratops in my garden, that would be exciting and almost certainly specious (pink elephants are another matter). It is the unexpected result that is exciting and gets hyped. One can say overhyped. There is pressure to get exciting results out quickly and widely distributed so you get the credit; a pressure to not check as carefully as one should, a pressure to ensure priority by not checking with one's peers.

Couple the low prior probability and the desire for fame with the ubiquity of error and you have the backdrop to most exciting new results being wrong. Not all exciting new results are wrong, of course. For example, the discovery of high temperature superconductors (high = liquid nitrogen temperature). This had crackpot written all over it. The highest temperature recorded earlier was 30deg Kelvin. But with high temperature superconductors, that jumped to 90 in 1986 and then shortly afterwards to 127deg kelvin. Surely something was wrong, but it wasn't and a Nobel Prize was awarded in 1987. The

---

[1] For more information on the examples, Google is your friend.

speed at which the Nobel Prize was awarded was also the subject of some disbelief. Why was the result accepted so quickly? Reproducibility. The results were made public and quickly and widely reproduced. It was not just the cronies of the discoverers who could reproduce them.

The lesson here is to distrust every exciting new science result: canals on Mars, n-ray, high temperature superconductors, faster than light propagation of neutrinos, the Higgs bosons and so on. Wait until they have been independently confirmed and then wait until they have been independently confirmed again. There is a pattern with these results that turn out to be wrong. In almost every example given above, the first attempts at reproducing the wrong results succeeded. People are in a hurry to get on the bandwagon; they want to be first to reproduce the results. But after the initial excitement fades, sober second thought kicks in. People have time to think about how to do the experiment better, time to be more careful. In the end, it is this third generation of experiments that finally tells the tale. Yeah, I know I should not have been sucked in by pentaquarks but they agreed with my preconceived ideas and the second generation experiments did see them in almost the right place. Damn. Oh well, I did get a widely cited paper out of it.

Once burnt, twice shy. So scientists become very leery of the next big thing. Here again, science is different than the law. In the law, there is a presumption of innocence until proven guilty. In other words, the prosecution must prove guilt; the suspect does not need to prove innocence. In science, the burden of proof is the other way around. The suspect—in this case, the exciting new result—must prove innocence. It is the duty of the proponents of the exciting new result to demonstrate the validity or usefulness of the new result. It is the duty of his peers to look for holes, because as the above examples indicate, they are frequently there.

53. CREATIVITY: AS IMPORTANT IN THE SCIENCES AS IN THE ARTS

In these essays I have discussed various aspects of the scientific method based on model building and testing against experiment. One aspect I have avoided until now is how the models are constructed. This is a logically distinct process from how models are tested. We have seen that models are tested by requiring them to make predictions that can be tested against observation. We can then rank models on parsimony and their ability to make successful predictions. But how are models constructed in the first place? Francis Bacon (1561 – 1626) and Isaac Newton (1642 – 1727) would have told you that models were deduced form observation by a method called induction. In this approach the construction and testing of models is seen as the same process, not two distinct processes. But induction is not valid and the making of models is, as noted, logically independent of their testing. This is a key point Karl Popper (1902 – 1994) made when he introduced the idea of falsification.

Let us turn to a master, Albert Einstein (1879 – 1955), and see what he says[1]:

> *The supreme task of the physicist is . . . the search for those most general, elementary laws from which the world picture is to be obtained through pure deduction. No logical path leads to these elementary laws; it is instead just the intuition that rests on an empathic understanding of experience. In this state of methodological uncertainty one can think that arbitrarily many, in themselves equally justified systems of theoretical principles were possible; and this opinion is, in principle, certainly correct. But the development of physics has shown that of all the conceivable theoretical constructions a single one has, at any given time, proved itself unconditionally superior to all others. No one who has really*

---

[1] As quoted in the Stanford Encyclopedia of Philosophy from MOTIVE DES FORSCHENS (1918).

gone deeply into the subject will deny that, in practice, the world of perceptions determines the theoretical system unambiguously, even though no logical path leads from the perceptions to the basic principles of the theory.

Very curious, *no logical path leads to these elementary laws.* Paul Feyerabend (1924 – 1994) would have agreed. He wrote a book entitled: AGAINST METHOD: OUTLINE OF AN ANARCHISTIC THEORY OF KNOWLEDGE. In this book he argued that there is no scientific method, but that knowledge advances through chaos. However, what Bacon and Newton missed, Feyerabend glimpsed through a haze and Einstein understood is that model construction is a creative, not algorithmic, process: *the intuition that rests on an empathic understanding of experience.* Science does not function by deducing models from observations. Rather we construct models and compare their predictions with what is observed. A falling apple inspired Newton, rising water in a bath inspired Archimedes, a dream inspired Kekulé (the structure of benzene). Or at least that is how the stories go. For model construction, Feyerabend is correct, anything goes—dreams, divine inspiration, pure luck, and especially hard work. Creativity by its very nature is chaotic and erratic.

We start with observations and say: what is the simplest model that could account for these observations? Once a model is constructed it is tested by the much more algorithmic or deterministic process of making predictions and checking them against observation. Now one of the necessary constraints in building scientific models is simplicity. Without simplicity we can get nowhere: an infinite set of models describe any finite set of measurements. That is why we cannot say: what model do these observations imply? There are infinitely many such models.

But simplicity comes at a price. For the sake of argument, take simplicity to be defined in terms of Kolmogorov complexity. This is a measure of the computational resources needed to specify the model. Now there is a theorem that says the Kolmogorov complexity cannot be determined algorithmically. If we accept the above identification of simplicity in scientific models with Kolmogorov complexity it then follows that scientific models cannot be constructed algorithmically from observations. So much for Francis Bacon, Newton and induction. The identification is probably not exact but nevertheless sufficiently close to reality to be indicative; model building cannot be algorithmic but rather is creative. Incidentally, ruling out algorithmically generated models implies that we cannot think of observations as axioms in an axiomatic model for knowledge.

The creative aspect of science is obscured by two things: the analytic aspect and the accumulative aspect. The analytic aspect of testing models tends to obscure the creativity in constructing them. We are so blinded by the dazzling mathematics virtuoso of a Newton that we fail to see how creative the development of his three laws was. Aristotle got it all wrong but Newton got it right; and that was through creativity not just math.

The accumulative nature of science gives a sense of inevitability that makes the creativity less obvious and the sense of creativity is lost in the noise of a thousand lesser persons. To quote Bertrand Russell[1]:

> *In science men have discovered an activity of the very highest value in which they are no longer, as in art, dependent for progress upon the appearance of continually greater genius, for in science the successors stand upon the shoulders of their predecessors; where one man of supreme genius has invented a method, a thousand lesser men can apply it. No transcendent ability is required in order to make useful discoveries in science; the edifice of science needs its masons, bricklayers, and common labourers as well as its foremen, master builders, and architects. In art nothing worth doing can be done without genius; in science even a very moderate capacity can contribute to a supreme achievement.*

While we common labourers may not be creative geniuses, the foremen, master builders and architects are. When it comes to creativity, Isaac Newton, Charles Darwin and Albert Einstein do not take a back seat to writers like William Shakespeare, Charles Dickens, or James

---

[1] From MYSTICISM AND LOGIC AND OTHER ESSAYS (1919), chapter II.

Joyce nor to painters like Michelangelo, Vincent van Gogh or Pablo Picasso.

54. HIGGS VERSUS DESCARTES: THIS ROUND TO HIGGS

René Descartes (1596 – 1650) was an outstanding physicist, mathematician and philosopher. In physics, he laid the groundwork for Isaac Newton's (1642 – 1727) laws of motion by pioneering work on the concept of inertia. In mathematics, he developed the foundations of analytic geometry, as illustrated by the term *Cartesian[1] coordinates*. However, it is in his role as a philosopher that he is best remembered, which is rather ironic, as his breakthrough method was a failure.

Descartes's goal in philosophy was to develop a firm foundation for all knowledge based on ideas that were so obvious they could not be doubted. His touchstone was that anything he perceived clearly and distinctly as being true was true. The archetypical example of this was the famous *I think therefore I am*. Unfortunately, little else is as obvious as that famous quote and even it can be—and has been— doubted.

Euclidean geometry provides the illusionary ideal to which Descartes and other philosophers have strived. You start with a few self-evident truths and derive a superstructure built on them. Unfortunately even Euclidean geometry fails that test. The infamous parallel postulate has been questioned since ancient times as being a bit suspicious. Other Euclidean postulates have also been questioned; extending a straight line depends on the space being continuous, unbounded and infinite.

So how are we to take Euclid's postulates and axioms? Perhaps we should follow the idea of Sir Karl Popper (1902 – 1994) and consider them to be bold hypotheses. This casts a different light on Euclid and his work; perhaps he was the first outstanding scientist. If we take his

---

[1] *Descartes* when Latinized is *Cartesius* hence *Cartesian*.

basic assumptions as empirical [1] rather than sure and certain knowledge, all we lose is the illusion of certainty. Euclidean geometry then becomes an empirically testable model for the geometry of space time. The theorems, derived from the basic assumption, are prediction that can be checked against observations satisfying Popper's demarcation criteria for science. Do the angles in a triangle add up to two right angles or not? If not, then one of the assumptions is false, probably the parallel line postulate.

Back to Descartes, who criticized Galileo Galilei (1564 – 1642) for having built *without having considered the first causes of nature, he has merely sought reasons for particular effects; and thus he has built without a foundation.* In the end, that lack of a foundation turned out to be less of a hindrance than Descartes's faulty one. To a large extent, science's lack of a foundation, such as Descartes wished to provide, has not been a significant obstacle to its advance.

Like Euclid, Sir Isaac Newton had his basic assumptions—the three laws of motion and the law of universal gravity—but he did not believe that they were self-evident; he believed that he had inferred them by the process of scientific induction. Unfortunately, scientific induction was as flawed as a foundation as the self-evident nature of the Euclidean postulates. Connecting the dots between a falling apple and the motion of the moon was an act of creative genius, a bold hypothesis, and not some algorithmic derivation from observation.

It is worth noting that, at the time, Newton's explanation had a strong competitor in Descartes's theory that planetary motion was due to vortices, large circulating bands of particles that keep the planets in place. Descartes's theory had the advantage that it lacked the *occult* action at a distance that is fundamental to Newton's law of universal gravitation. In spite of that, today, Descartes's vortices are as unknown as his claim that the pineal gland is the seat of the soul; so much for what he perceived clearly and distinctly as being true.

Galileo's approach of solving problems one at time and not trying to solve all problems at once has paid big dividends. It has allowed sci-

---

[1] As in the final analysis they are.

ence to advance one step at a time while Descartes's approach has faded away as failed attempt followed failed attempt. We still do not have a grand theory of everything built on an unshakable foundation and probably never will. Rather we have models of widespread utility. Surely that is enough, even if they are built on shaky foundations.

Peter Higgs (b. 1929) follows in the tradition of Galileo. He has not, despite his Noble prize, succeeded, where Descartes failed, in producing a foundation for all knowledge; but through creativity, he has proposed a bold hypothesis whose implications have been empirically confirmed. Descartes would probably claim that he has merely sought reasons for a particular effect: mass. The answer to the ultimate question about life, the universe and everything still remains unanswered, much to Descartes's chagrin but as scientists we are satisfied to solve one problem at a time then move on to the next one.

55. THE ACCUMULATIVE NATURE OF SCIENCE

Is science accumulative? Is the Pope a Catholic? Some things are truly self-evident. The accumulative nature of science is one of them. But the different ways science is accumulative does hold some surprises. Consider home improvement. We can add onto the top, build out from the side, fix the broken window, or build down from the foundations. Science is accumulative in all these directions. Think of classical mechanics and planetary motion. After Isaac Newton (1642 – 1727) introduced his three law of motion, various people, most notably Joseph-Louis Lagrange (1736 – 1813) and William Hamilton (1805 – 1865), developed more mathematically sophisticated treatments of classical motion. They used Newton's work as a starting point and built new stories onto the superstructure. Pierre-Simon Laplace (1749 – 1827) added onto Newton's work in other ways. He found an error in Newton's calculation of planetary stability and added the nebular hypothesis to describe the origin of the solar system. The first of these corrected the structure Newton had built—replaced the broken window if you like—while the second, the nebular hypothesis, added a new room on the side. It extended Newton's ideas beyond where they were originally applied. The discovery of Uranus by William Herschel (1738 – 1822) can also be considered a sidewise extension to planetary system; the discovery left the original work intact but extended it outward.

These advances all left the paradigm of classical mechanics intact but built on the foundation Newton had laid. But quantum mechanics was a whole different story. It left the superstructure intact but changed the foundation; like the magician's trick of pulling the tablecloth off the table while leaving the dishes in place. The advent of quantum mechanics did not require the recalculation of planetary orbits. The work of Newton, Laplace, Lagrange, and Hamilton could still be applied as before but only to a fixed range of phenomena. Quantum mechanics kept all the successes of classical mechanics, but put it on a new foundation.

Now, quantum mechanics frequently is seen as a complete overthrow of classical mechanics and if you are looking at the metaphysics, that is true. However, no one should take metaphysics seriously anyway. From the point of view of the person calculating planetary orbits, nothing changed when Schrodinger introduced his eponymous equation. Schrodinger built on the work of Hamilton just as much as Hamilton built on the work of Newton. (Quantum mechanics is built on Hamilton's formulation of classical mechanics.) Whereas Hamilton added to the superstructure, Schrodinger helped replace the foundation. Both added to the existing structure rather than demolishing it, and the smoke went up the chimney just the same[1]. Or rather, the planets went round the sun just the same.

Replacing the foundation is largely synonymous with Thomas Kuhn's (1922 – 1996) idea of paradigm change. This is the reductionist's dream and the foundations in various fields of science are indeed frequently replaced: quantum gravity will replace quantum field theory, which replaced quantum mechanics, which in turn replaced classical mechanics. But only the foundation was replaced, the superstructure was left intact. A similar process happened with this sequence: indivisible atoms, atom structure, nuclear structure, nucleon structure, and the standard model.

---

[1] From a children's song by Fred Chandler, 1901.

Thus, we see how science advances: fixing errors (Laplace), refining formalisms (Hamilton, Lagrange), extending to new areas (Laplace, Herschel), and replacing the foundations (Schrodinger). But these are all extensions to the existing knowledge. When we forget this, mistakes are made. When quantum chromodynamics was introduced, it changed the foundation of nuclear physics, but left most of the previous understanding of nuclear physics intact. The overzealous proponents of the quantum chromodynamics did not understand this and claimed that nuclear physics would have to be largely redone. But that was nonsense; we made a few minor changes and carried on. Science is amazing in that it can easily change the foundation without major damage to the superstructure. Try doing that with a sky scraper or even a two-storey house.

The reason this all works is that science is modular, with fairly well-defined interfaces between the models. Consider chemistry. At one side, quantum chemistry is closely related to physics and shares a common formalism: quantum mechanics. In the middle, chemistry developed independently of physics and did not depend on the quantum chemistry foundation. But then, parts of biology and applied science use chemistry as a foundation to build on; interlinked but each progressing separately.

So science oozes onwards in all directions: upward, downward, sideways, and inward. It discards what is no longer useful—yet, for the most part, the older models provide the scaffolding to support the new, and the more recent insights are obtained without destroying the older ones. And science unfolds as it should, building knowledge one room at a time.

56. THE RAVEN PARADOX AND THE FLAW IN VERIFICATION

One of the more interesting little conundrums in understanding science is the raven paradox. It was proposed by Carl Hempel (1905 – 1997) in the 1940s. Consider the statement: *all ravens are black*. In strict logical terms, this statement is equivalent to: *everything that is not black is not a raven*. To verify the first we look for ravens that are black. To verify the latter we look for coloured objects that are not ravens. Thus finding a red (not black) apple (not raven) confirms that: *everything that is not black is not a raven*, and hence that: *all ravens are black*. It seems strange, to learn about the colour of birds, we study a basket of fruit.

While the two statements may be equivalent for ravens, they are not equivalent for snarks. The statement: *everything that is not black is not a snark*, is trivially true since snarks do not exist, except in Lewis Carroll's imagination. However, the statement: *all snarks are black*, is rather meaningless since snarks of any colour do not exist (boojums are another matter). Hence, the equivalence of the two statements in the first paragraph relies on the hypothesis that ravens do exist.

One resolution of the paradox is referred to as the Bayesian solution. The ratio of ravens to non-black objects is so near to zero that it makes no difference. Thus finding 20 black ravens is more significant than find 20 non-black, non-ravens. You have sampled a much larger fraction of the objects of interest. While it is not possible to check a significant fraction of non-black objects in the universe, it may be possible to check a significant faction of ravens, at least those which are currently alive.

But the real solution to the problem seems to me to lie in different direction. Finding a red apple confirms not only that all ravens are black but also that all ravens are green, or chartreuse, or even my daughter's favorite colour, pink. The problem is that a given observation can confirm or support many different, and possibly contradictory, models. What we do in science is compare models and see which is better. We grade on a relative, not absolute scale. To quote Sir Carl Popper (1902 – 1994):

And we have learnt not to be disappointed any longer if our scientific theories are overthrown; for we can, in most cases, determine with great confidence which of any two theories is the better one. We can therefore know that we are making progress; and it is this knowledge that to most of us atones for the loss of the illusion of finality and certainty.

We do not want to know if: *all ravens are black* is true but rather if the statement *all ravens are black* is more accurate than the statement *all ravens are green*. A red apple confirms both statements, while a green apple confirms one and is neutral about the other. Thus the relative validity of the two statements cannot be checked by studying apples, but only by studying ravens to see what colour they are. Thus, the idea of comparing models leads to the intuitive result. Whereas, thinking in terms of absolute validity leads to nonsense: here, check this stone to see if ravens are black. Crack, tinkle (sound of broken glass as stone misses raven and goes through neighbor's window).

We can go farther. Consider the two statements: *all ravens are black*, and *some ravens are not black*. The relative validity of these two statements cannot be checked by studying apples or even black ravens. Rather what is needed is a non-black raven. This is just the idea of falsification. Hence, falsification is just a special case of comparing models: A is correct, A is not correct.

In practice, all ravens are not black. There are purported instances of white ravens. Google says so and Google is never wrong. Right? Thus, we have the statement: *most ravens are black*. This statement does not imply anything about non-black objects; they may or may not be ravens. Curious... this whole raven paradox was based on a false statement and as with: *all ravens are black*, most absolute statements are false, or at least, not known for certain.

Even non-absolute statements can lead to trouble. Consider: *most ravens are black*, and: *most ravens are green*. So we merrily check ravens to see which is correct. But is it not possible that the green ravens blend in so well with the green foliage that we are not aware that they are there? Rather like the elephants in the kid's joke that paint their toenails red so they can hide in cherry trees. Works like a charm. Who has seen an elephant in a cherry tree? We are back to the Duhem-Quine thesis that no idea can be checked in isolation. Ugh. So, why do we dismiss the idea of perfectly camouflaged green ravens and red-toenailed elephants? Like any good conspiracy theory, they can only be eliminated by an appeal to simplicity. We eliminate the perfectly camouflaged green raven by parsimony, and as for the red apple, I ate it for lunch.

57. HIGGS VS POPPER: FALSIFICATION FALSIFIED

Finding the Higgs boson had no epistemic value whatsoever. That is certainly a provocative statement. However, if you believe that science is defined by falsification, it is a true one. Can it really be true, or is the flaw in the idea of falsification? Should we thumb our noses at Karl Popper (1902 – 1994), the philosopher who introduced the idea of falsification?

The Higgs boson, the last remaining piece of the standard model, has been the object of an enormous search involving scientists from around the world. The ATLAS collaboration alone has 3000 participants from 174 institutions in 38 different countries. Can only the failure of this search be significant? Should we have sent out condolence letters when the Higgs boson was found? Were the Nobel prizes for the W and Z bosons a mistake?

Imre Lakatos (1922 – 1974), a neo-falsificationist and follower of Popper, states it very cleanly and emphatically[1]:

> *But, as many sceptics pointed out, rival theories are always indefinitely many and therefore the **proving** power of experiment vanishes. One cannot learn from experience about the truth of any scientific theory, only at best about its falsehood: **confirming instances have no epistemic value whatsoever** [emphasis in the original].*

Yikes! What is going on? Can this actually be true? No! To see the flaw in Lakatos's argument, let's consider an avian metaphor—this time *Cygnus* not *Corvus*. Consider the statement: *All swans are white.* (Here we go again.) Before 1492, Europeans would have considered this a valid statement. All the swans they had seen were white. Then Europeans started exploring North America. Again, the swans were white. Then they went on to South America and found swans with black necks (*Cygnus melancoryphus*) and finally to Australia where

---

[1] From THE ROLE OF CRUCIAL EXPERIMENTS IN SCIENCE, STUD. HIST. PHIL. SCI. 4 (1974).

the swans are black (*Cygnus atratus*). By the standards of the falsificationist, nothing was learned when white swans were found, but only when the black swans or partially black swans were found. With all due respect, or lack of same, that is nonsense. It is the same old problem: you ask a stupid question you get a stupid answer. Did we learn anything when white swans were found in North America? Yes. We learned that there were swans in North America and that they were white. Based on having white swans in Europe, we could not deduce the colour of swans in North America or even that they existed. In Australia, we learned that swans existed there and were black. Thus, we learned a similar amount of information in both cases— really nothing more or nothing less. The useful question is not, *Are all swans white?* Rather, *On which continents do swans exist and what color are they on each continent?*

Moving on from birds to model cars (after all, the standard model of particle physics is a model). What can we learn about a model car? Certainly, not if it is correct. Models are never an exact reproduction of reality. But, we can ask, *Which part of the car is correctly described by the model? Is it the color? Is it the shape of the head lights or bumper?* The same type of question applies to models in science. The question is not, *Is the standard model of particle physics correct?* We knew from its inception that it is not the answer to the ultimate question about life, the universe and everything. The answer to that is 42 (Deep Thought, from THE HITCHHIKER'S GUIDE TO THE GALAXY by Douglas Adams (1952 – 2001)). We also know that the standard model is incomplete because it does not include gravity. Thus, the question never was, *Is this model correct?* Rather, *What range of phenomena does it usefully describe?* As C.I. Lewis (1883 – 1964) expressed it more formally: *The problem now becomes whether, or how far, this deductive system is applicable to empirical facts*[1]. The Standard Model has a long history of successful predictions and is applicable to a lot of empirical observations. So, like the model car, it captures some aspects of reality, but not all.

---

[1] THE MIND AND THE WORLD ORDER, 1929

Finding the Higgs boson helps define what part of reality the standard model describes. It tells us that the standard model still describes reality at the energy scale corresponding to the mass of the Higgs boson. But, it also tells us more: It tells us that the mechanism for electroweak symmetry breaking—a fundamental part of the model—is adequately described by the mechanism that Peter Higgs (b. 1929) and others proposed and not some more complex and exotic mechanism.

The quote from Lakatos, given above, misses a very important aspect of science—parsimony. The ambiguity noted there is eliminated by the appeal to simplicity. The standard model of particle physics describes a wide range of experimental observations. Philosophers call this phenomenological adequacy. But a lot of other models are also phenomenologically adequate. The literature is filled with extensions to the standard model that agree with the standard model where the standard model has been experimentally tested. They disagree elsewhere, usually at higher energy. Why do we prefer the standard model to these pretenders? Simplicity and only simplicity. And the standard model will reign supreme until one of the more complicated pretenders is demonstrated to be more phenomenologically adequate. In the meantime, I will be a heretic and proclaim that the finding of the Higgs boson does indeed confirm the standard model. All the hoopla when the Higgs boson was found, celebrating a confirmed prediction, puts paid to the idea that science is only about falsification. Falsification has been falsified.

58. THE ANTHROPIC PRINCIPLE OR THE AGROSTIC PRINCIPLE?

In honour of spring (or April first).

As I drive to and from work in Vancouver, I notice that even in winter, the grass is green. In the spring, people are out fertilizing their lawns and in summer watering them (even when they are not allowed to)—mollycoddled grass! They are now even putting grass on the top of buildings. You would almost think that Vancouver exists for the benefit of grass. But it is not just Vancouver; we have wide areas of the world devoted to grass, from bamboo to grain. You would think the world was created for the benefit of grass. After all, the earth is just the right distance from the sun to allow grass to flourish. Farther from the sun, it would be cold and arid like Mars. Closer to the sun it would be hot and sterile like Venus. Thus, we have what is known in the trade as the agrostic[1] principle: the philosophical argument that observations of the physical universe must be compatible with the preferred status of grass.

As just mentioned, the earth is just at the right distance from the sun for grass to flourish. But it goes beyond that. Carbon is a major component of grass. However, the creation of carbon in stars depends critically on the existence of an excited state in carbon, known as the Hoyle state, with exactly the right energy. If that state were not there, there would be no carbon and hence no grass. The horror of it! Just think, no grass. And it all depends on having the nuclear state at just the right energy.

The Hoyle state is not the only coincidence necessary for the existence of grass. If the fundamental constants of nature, things like the fine structure constant or the gravitational constant (big G) were slightly different, the universe would not support the existence of grass. There are two solutions to this problem. One is to assume that there is an intelligent designer with an inordinate fondness for grass who fine-tuned the universe so grass could exist. Now, there is a minority opinion[2] that it is not grass that he is fond of, but rather beetles (coleoptera) and that he only created grass as a source of feed for beetles. After all, there is the order of a million species of beetles. But as I just said, the coleopteric principle is distinctly a minority position, but we should be open minded. The other explanation of the fine tuning of the universe is based on the idea of the multiverse. This is the idea that many different universes exist with all possible values of the physical constants and that we are in the one in which grass is possible. Again, note the preferred role of grass. The evidence for this scenario, at the present time, is no stronger than that for the existence of the coleopterophillic intelligent designer.

---

[1] From the Greek word ἄγρωστις for grass.
[2] Probably held by zoologists.

Now one might ask what role consciousness and intelligence have in all this. The answer to that is fairly self-evident. The main role of consciousness and intelligence is the development of civilization, and the main role of civilization is the development of agriculture. It should be obvious to even the most obtuse reader that the main purpose of agriculture is to permit grass to more effectively compete with trees. Just think of the extent to which farmers have replaced forests with grassland. The bringing of European *civilization* to North America had as its main effect, the replacement of forest with grassland. It had some unfortunate side effects, like the creation of the United States of America, but what is more important—people or grass?

As further evidence of the agrostic principle, I note that it provides the only possible explanation for the existence of golf courses and cricket pitches. The very idea of grown men or women hitting a ball with a club to prove their virility is silly. Now artificial turf may be considered as evidence against the agrostic principle, but artificial turf seems to be a passing fad. In just 13 years, between 1992 and 2005, the National Baseball League went from having half of its teams (6 of 12) using artificial turf to all of them—now up to 16— playing on natural grass. As for football (soccer), artificial turf is widely banned except for women's competitions. Enough said.

The agrostic principle also highlights flaws in ancient Greek philosophy. Plato believed that the GOOD was contemplating his ideals or ideas. That is incorrect; the greatest good is cultivating and contemplating grass. Like Euclid's postulates, that should be self-evident. That the smoking of grass is the greatest good is a corruption of Epicurus's teaching. Rather, he was the first of the new atheists. The Sophists, on the other hand, where the first postmodernists and believed that it was impossible to decide if contemplating or smoking grass was the greatest good. After smoking a few joints, the latter is probably true. Socrates believed that nothing could be learned from nature. Perhaps if he had spent more time cultivating and contemplating grass, he would not have been compelled to drink hemlock. However, Aristotle may have been onto something with his final cause or teleology. Evolution shows its bareness by failing to recognize that consciousness and intelligence arose due to the teleological purpose (final cause) of helping grass compete with trees. This is probably the best example of the need for Aristotle's final cause that can be found in nature. Unfortunately, Aristotle starting worrying about essences rather than cultivating and contemplating grass. Thus, the Greek civilization decayed. And my wife wants me to replace the lawn with a garden. The end of western civilization is in sight.

The agrostic principle has some naysayers. Douglas Adams (1952 – 2001) gives the example in his HITCHHIKER'S GUIDE TO THE GALAXY of the puddle which observed how well it fitted the hole it was in and concluded that the hole and the universe where created expressly for its benefit. It was consequently quite surprised and distressed when it evaporated. Imagine; the gall of Adams using satire to attack the agrostic principle. Now, of course, the properties of the hole can be deduced from the properties of the puddle, but this should not be used to infer the universe was not created for the sole benefit of the puddle. Some people have followed the example of Adams's puddle and claimed that since humans nicely fit a hole in the universe, the universe was created for their benefit (this is sometimes call the anthropic principle). These people will probably be surprised when humans become extinct. The superiority of the agrostic principle to the anthropic principle is shown by the observation that while *homo sapiens* have existed for about 200,000 years, grass tickled the feet of dinosaurs over sixty million years ago. And grass will probably still exist after humans have, through sheer stupidly, destroyed themselves and have been replaced by a group with less intelligence and more wisdom, perhaps the *coleoptera*.

59. COSMIC INFLATION OR ARE THE PHOTONS MESSING WITH OUR MINDS?

Another April first post.

Recently it was announced that a smoking gun has been found for cosmic inflation but could it be instead the smoking[1] gun for a grand

---

[1] And no, I have not been smoking with the former mayor of a certain large Canadian city.

conspiracy, the mother of all conspiracies, the conspiracy theory to put all other conspiracy theories to shame? You may have your favourite conspiracy theory: the Roswell cover up, who shot JFK, the suppression of perpetual motion machines by the energy companies or the attempt by communication departments to take over the world. As a professor, once said: *Just because you are paranoid does not mean they are not out to get you.* (Followed closely by my second-best piece of advice, *never trust a communications expert.*) But the conspiracy I am talking about is on a much larger scale, a cosmic scale, the conspiracy of all elementary particles in the universe to use their free will for the sole purpose of messing with the minds of humans and particularly that subclass of humans known as particle physicists[1]. The professor was right to be paranoid.

We hold these truths to be self-evident, that all particles are created equal and endowed by their Creator with free will. Surely you do not question that elementary particles have free will. Consider the muon. It decays at a time if its own choosing. There is no rule that says when a given muon will decay. It is decided by the muon in its own stubborn way. But you still claim that only humans have free will. Why? Could it not be just of part the grand conspiracy of elementary particle to give humans the illusion of free will? Besides we all know politicians do not have free will. They just do, as a Pavlovian reflex, what they think will get them the most votes. Hence the mess the world is in. But I digress.

To explore further: what criteria is there to decide if a given object has free will? Is it just unpredictability? If so, then Vancouver weather has free will. We have already decided that politicians do not have free will. But what about dogs? Are all their reactions Pavlovian conditioning? Most certainly not. Hence they are better candidates to have free will than politicians. And what about cats? Cats certainly have will but it is free? Does the fact cats cannot be herded indicate they have free will? And cows, do cows have free will? Does the fact they can be herded indicate they do not have free will? Or that plant on my window sill beckoning to be watered? Does it cause its leaves

---

[1] Nuclear physicists have always known that something was messing with the minds of particle physicists.

155

to droop out of the exercise of its own free will just to annoy me? Possibly (there it goes again). It seems to me the hallmark of free will is precisely that combination of random and non-random behaviour found in elementary particles and not in politicians. Hence another interpretation of quantum uncertainty, it is just elementary particles exercising their free will.

You may be surprised that I claimed all particles are created equal. Surely the neutrino and electron have different properties and are not equal. But that is just them exercising their free will. Those particles we call electrons have decided (note the verb decided) to behave as if they had a given set of interactions while the neutrinos have decided to be behave like they have a different set of interactions. The interactions, themselves, are illusionary created by the particles exercising their free will. There was probably a grand council meeting at the beginning of time where the grand conspiracy was initiated and the roles assigned to different sets of particles. Indeed, it may have been that grand council meeting that started time itself.

The free will of particles also explains the problem of evil. From earthquakes to global warming[1], the evil consequences are due to particles exercising their free will. This type of evil could only be eliminated by denying particles free will—an even greater evil.

Now back to the initial observation about the polarization of photons in the cosmic microwave background. Surely that polarization is not due to gravitational waves but due to photons exercising their free will to mislead humans. Is it not more reasonable to assume that the measured polarization[2] is due to free will than to some far distant interaction with gravitons? (Do gravitons even exist? They have never been directly observed.)

This conspiracy theory—just like all conspiracy theories—accounts for all the known facts and cannot be disproved. It therefore must be

---

[1] Global warming is due to particles excising their free will, not to carbon dioxide emissions.
[2] But beware the Jennings principle: Most exciting new results are wrong.

correct and we have shown conclusively that it is particles, not people, that have free will and the photons are trying to mess with our minds[1].

60. STRING THEORY AND THE SCIENTIFIC METHOD

It seems some disagreements are interminable: the Anabaptists versus the Calvinists, capitalism versus communism, the Hatfields versus the McCoys, or string theorists versus their detractors. It is the latter I will discuss here although the former may be more interesting. This essay is motivated by a comment in the December 16, 2014 issue of Nature by George Ellis and Joe Silk. The comment takes issue with attempts by some string theorists and cosmologists to redefine the scientific method by eliminating the need for experimental testing and relying on elegance or similar criteria instead. I have a lot of sympathy with Ellis and Silk's point of view but believe that it is up to scientists to define what science is and that hoping for deliverance by outside people, like philosophers, is doomed to failure.

To understand what science is and what science is not, we need a well-defined model for how science behaves. Providing that well-defined model is the motivation behind each of my essays. The scientific method is quite simple: build models of how the universe works based on observation and simplicity. Then test them by comparing their predictions against new observation. Simplicity is needed since observations underdetermine the models (see for example: Willard Quine's (1908 – 2000) essay: The Two Dogmas of Empiricism). Note also that what we do is build models: the standard model of particle physics, the nuclear shell model, string theory, etc. Quine refers to the internals of the models as myths and fictions. Henri Poincaré (1854 – 1912) talks of conventions and Hans Vaihinger (1852 –1933) of the *philosophy of as if* otherwise known as fictionalism. Thus it is important to remember that our models, even the so-called theory of everything, are only models and not reality.

---

[1] This is not the perfect particle physics essay because it does not mention the Higgs boson, the LHC or super symmetry. Oops, maybe now it is.

It is the feedback loop of observation, model building and testing against new observation that define science and give it its successes. Let me repeat: The feedback loop is essential. To see why, consider the example of astrology and why scientists reject it. Its practitioners consider it to be the very essence of elegance. Astrology uses careful measurements of current planetary locations and mathematics to predict their future locations, but it is based on an epistemology that places more reliance on the eloquence of ancient wisdom than on observation. Hence there is no attempt to test astrological predictions against observations. That would go against their fundamental principles of eloquence and the superiority of received knowledge to observation. Just as well, since astrological predictions routinely fail. Astrology's failures provide a warning to those who wish to replace prediction and simplicity with other criteria. The testing of predictions against observation and simplicity are hard taskmasters and it would be nice to escape their tyranny but that path is fraught with danger, as astrology illustrates. The feedback loop from science has even been picked up by the business management community and has been built into the very structure of the management standards (see ISO Annex SL for example). It would be shame if management became more scientific than physics.

But back to string theory. Gravity has always been a tough nut to crack. Isaac Newton (1643 – 1727) proposed the decidedly inelegant idea of instantaneous action at a distance and it served well until 1905 and the development of the special theory of relativity. Newton's theory of gravity and special relativity are inconsistent since the latter rules out instantaneous action at a distance. In 1916, Albert Einstein (1879 – 1955) with an honorable mention to David Hilbert (1862 – 1943) proposed the general theory of relativity to solve the problem. In 1919, the prediction of the general theory of relativity for the bending of light by the sun was confirmed by an observation by Arthur Eddington (1882 – 1944). Notice the progression: conflict between two models, proposed solution, confirmed prediction, and then acceptance.

Like special relativity and Newtonian gravity, general relativity and quantum mechanics are incompatible with one another. This has led to attempts to generate a combined theory. Currently string theory is the most popular candidate, but it seems to be stuck at the stage general relativity was in 1917 or maybe even 1915: a complicated (some would say elegant, others messy) mathematical theory but unconfirmed by experiment. Although progress is definitely being made, string theory may stay where it is for a long time. The problem is that the natural scale of quantum gravity is the Planck mass and this scale is beyond what we can explore directly by experiment. However, there is one place that quantum gravity may have left observable traces and that is in its role in the early Universe. There are experimental hints that may indicate a signature in the cosmic microwave background radiation but we must await further experimental results. In the meantime, we must accept that current theories of quantum gravity are doubly uncertain. Uncertain, in the first instance, because, like all scientific models, they may be rendered obsolete by new a understanding and uncertain, in the second instance, because they have not been experimentally verified through testable predictions.

Let's now turn to the question of multiverses. This is an even worse dog's breakfast than quantum gravity. The underlying problem is the fine tuning of the fundamental constants needed in order for life as we know it to exist. What is needed for life, as we do not know it, to exist is unknown. There are two popular ideas for why the Universe is fine-tuned. One is that the constants were fine-tuned by an intelligent designer to allow for life as we know it. This explanation has the problem that by itself it can explain anything but predict nothing. An alternate is that there are many possible universes, all existing, and we are simply in the one where we can exist. This explanation has the problem that by itself it can explain anything but predict nothing. It is ironic that to avoid an intelligent designer, a solution based on an equally dubious just so story is proposed. Since we are into just so stories, perhaps we can compromise by having the intelligent designer choosing one of the multiverses as the one true Universe. This leaves the question of who the one true intelligent designer is. As an old farm boy, I find the idea that Audhumbla, the cow of the Norse creation myth, is the intelligent designer to be the most elegant. Besides, the idea of elegance, as a deciding criterion in science, has a certain bovine aspect to it. Who decides what constitutes elegance? Everyone thinks their own creation is the most elegant. This is only possible in Lake Wobegon, where all the women are strong, all the men are good-looking, and all the children are above average (A PRAIRIE HOME COMPANION, Garrison Keillor (b. 1942)). Not being in Lake Wobegon, we need objective criteria for what constitutes elegance. Good luck with that one.

Some may think the discussion in the last paragraph is frivolous, and quite by design it is. This is to illustrate the point that once we allow the quest for knowledge to escape from the rigors of the scientific method's feedback loop all bets are off and we have no objective reason to rule out astrology or even the very elegant Audhumbla. However, the idea of an intelligent designer or multiverses can still be saved if they are an essential part of a model with a track record of successful predictions. For example, if that animal I see in my lane is Fenrir, the great gray wolf, and not just a passing coyote, then the odds swing in favor of Audhumbla as the intelligent designer and Ragnarok is not far off. More likely, evidence will eventually be found in the cosmic microwave background or elsewhere for some variant of quantum gravity. Until then, patience (on both sides) is a virtue.

> *Though the mills of science grind slowly;*
> *Yet they grind exceeding small;*
> *Though with patience they stand waiting,*
> *With exactness grind they all.* [1]

61. THE ROLE OF THE CRUCIAL EXPERIMENT

The idea of a crucial experiment that decisively confirms a model goes back at least to Francis Bacon (1561 – 1626) who used the term *instantia cruci*. Later, the term *experimentum crucis* was coined by Robert Hooke (1635 – 1703) and used by Isaac Newton (1642 – 1727), in particular with regard to his theory of light. Alternatively, Pierre Duhem (1861 – 1916) strongly disagreed with the possibility of crucial experiments. Somewhat in anticipation of Thomas Kuhn's (1922 – 1996) paradigms, Duhem realized that scientific theories or models do not stand alone, but rather come coupled with auxiliary

---

[1] With apologies to Henry Wadsworth Longfellow (1807 – 1882).

assumptions. Was what Galileo (1564 – 1642) saw through the telescope features of the heavens, or only of his telescope, as some of his detractors claimed? One has to consider the combined heavens-telescope system to decide. When the detector is as complex as the ATLAS detector at CERN the question is even more apropos.

Karl Popper (1902 – 1994) refined the idea of the crucial experiment to one that falsifies a given model. But the Duhem-Quine hypothesis, a variation of Duhem's idea, makes the point that falsification, at least in its naïve form, falls victim to the same holistic argument: we can never test a single model in isolation. So is the idea of a crucial experiment just a will-o-the-wisp that vanishes on more careful evaluation?

We can think of many examples: Sir Arthur Eddington's measurement of the bending of star light by the sun, the discovery of high-temperature superconductors, the measurement of the three degree microwave background, the Michelson-Morley experiment, and so on. Did none of these play a critical role in the history of science? I would suggest they did, but not in the simple manner suggested by Bacon or Popper.

Consider the Michelson-Morley experiment of 1887 which tried and failed to measure the motion of the earth relative to the ether. Scientists did not do a Chicken Little impersonation and run around claiming the sky was falling or, in this case, that Newton (Newton's laws of motion) and Maxwell (electromagnetism) were wrong. Rather, they started trying to understand what the explanation could be. This led to ideas like ether drag (the earth entraining the ether) or Lorentz-Fitzgerald contraction (the idea that objects shorten in the direction of motion). The latter idea was developed and expanded upon by Lorentz and Poincaré who developed the math for special relativity. Einstein claimed he was unaware of the Michelson-Morley experiment, but he was certainly aware of Lorentz's early attempts to understand that experiment. Thus, the Michelson-Morley experiment started a chain of events that inexorably led to special relativity, not in one easy step, but eventually and inevitably. If special relativity had been proposed thirty years sooner, it would have been treated as a curiosity like the Copernicus model when it was first proposed.

As another example, consider the measurement of the bending of light by the sun. The general theory of relativity and classical mechanics differ by a factor of two. Eddington's 1919 experiment gave a result closer to general relativity and hence contributed to the early acceptance of general relativity (not that people are not still trying to test it; that is as it should be). A more striking example was the discovery of the three-degree kelvin cosmic microwave background. Before then, there were two models for the universe, both with strong support: the steady-state model and the big bang model. While the microwave background was a big boost for the big bang model, the steady-state model did not give up without a struggle. There were various attempts to describe the microwave background in the steady-state model but they were too little too late. Like the Michelson-Morley experiment, the discovery of the microwave background started a chain reaction that led to the acceptance of one model and the rejection of another.

Perhaps the best way of thinking of crucial experiments is not that they prove (that ugly word) one model better than another, but that they serve as a catalyst. Or perhaps, one can think of a super-cooled fluid that when slightly disturbed, suddenly solidifies. The same phenomenon is seen with people. A group is sitting at lunch and when one gets up to go they all go, but only if the circumstances are right. Consider the discovery of the J/Ψ particle. The time was right and the background had been prepared so that when it was discovered, the particle physics community solidified around the quark model. Similarly, you can consider Galileo turning his telescope on the heavens as providing the catalyst for the acceptance of the heliocentric model.

Like models, experimental results do not exist in isolation. Rather, they build on each other and are given meaning by the prevailing models. The role of crucial experiments should be seen in relation to that milieu. They do not singlehandedly overturn or confirm the status quo, but rather, start chains of events that lead to or act as tipping points for the establishment of new paradigms. Thus, crucial experiments do exist but not in the naïve manner envisioned by Bacon, Hooke, or Popper.

62. THE ARGUMENT FROM DESIGN

Central to the scientific method is a process for deciding between conflicting models of how the universe operates. It is very instructive to apply this process to the argument from design for the existence of a higher intelligence in the universe. The argument from design is commonly associated with William Paley (1743 – 1805) and for those who like big words, is also called the teleological argument for God's existence. A counter argument is given in Richard Dawkins's book: THE BLIND WATCH MAKER. The basic argument from design is, however, much older than Paley; it goes back to the ancient Greeks. Needless to say, Dawkins's book has failed to lay the argument to rest. If one checks the current state of the arguments on the topic[1], they typically are of the form: *Anyone who does not recognize design in the universe is in denial,* and the counter argument is: *Those who see design in the universe are delusional.* Needless to say, neither argument is particularly convincing. So what can the scientific method add to resolving the impasse? Quite a bit actually.

Let's begin by looking at the actual form of the argument. It was stated succinctly by Cicero (106 BCE – 43 BCE) (quoted from the WIKIPEDIA):

> *When you see a sundial or a water-clock, you see that it tells the time by design and not by chance. How then can you imagine that the universe as a whole is devoid of purpose and intelligence, when it embraces everything, including these artifacts themselves and their artificers?*

This analogy was expanded upon, most famously, by Paley (quoted from the WIKIPEDIA):

> *[S]uppose I found a watch upon the ground, and it should be inquired how the watch happened to be in that place, I should hardly think ... that, for anything I knew, the watch might have always been there. Yet why should not this an-*

---

*swer serve for the watch as well as for [a] stone [that hap-
pened to be lying on the ground]? ... For this reason, and for
no other; namely, that, if the different parts had been differently shaped from what they are, if a different size from what
they are, or placed after any other manner, or in any order
than that in which they are placed, either no motion at all
would have been carried on in the machine, or none which
would have answered the use that is now served by it.*

So what about the watch and how do we know that it was designed? We begin with one of the mantras of this series of essays: THE MEANING IS IN THE MODEL. To understand the watch and its creation, our mind, either consciously or unconsciously, develops a model for its origin. The watch is deduced to have been made by humans, not by non-human agencies, and humans do things by design. Thus, by a two-step process we arrive at design. Now, the watch is fairly obvious, but what about that pointed rock on the ground? Is it due to design or natural causes? Is it simply a broken rock or is it an arrow head? Here the question of design is strictly one of whether it was made by humans or not. If the indications on the rock show signs of human manufacture it is considered due to design, and if not, then accident.

The typical theist would claim that the universe and everything in it is designed. Thus, we cannot do the comparison of something designed to something that was not designed; a technique that was useful in deciding if the watch was humanly designed or not. So how do we tell if something is designed or not? Use the methodology from science, of course.

In science, there are two distinct steps with any model: first the model must be constructed, and then it must be tested. Model construction is a creative activity and does rely on analogy and pattern recognition. Thus, in the initial stage, the argument from design is on good grounds. Now for the crux of the matter: the crucial test is neither how good the analogy is, nor how striking the apparent pattern, but rather if the argument from design passes the tests of parsimony and also makes successful predictions for observations. The scientific method defines three criteria for judging models: the successful description of past observations, the ability to make correct predictions for future observations, and simplicity. Being able to describe past observations is just the price to play the game, and with sufficient ingenuity, can usually be done. The definitive test of a scientific model is the ability to make predictions for novel phenomena. By predictions, I mean definite predictions that can be falsified. Not the kind of predictions made by Nostradamus that after the fact can be claimed to have been fulfilled, but rather definite predictions that can be tested, like it will rain tomorrow at TRIUMF between 3:00 and 4:00 pm.

Finally, there is simplicity. Yes, there is always simplicity or parsimony. By simplicity, I mean the elimination of assumptions that do not help the model make predictions. Today, common descent for living things is pretty much established and is mainly challenged by gross violations of the simplicity principle. A prime example is the omphalos hypothesis of Phillip Gosse (1810 – 1888). He stated that the world was created six thousand years ago, but in a manner that cannot be distinguished from one that is much older. As pointed out in a previous essay, that hypothesis can only be eliminated by an appeal to parsimony. As for design, natural selection is one way of generating the design of living things without the need for external intelligence and, at least at the small scale, natural selection is observed to be happening. So, can an external intelligence as suggested by the argument from design, or the idea of intelligent design, add anything useful to this? Or can they both be eliminated, like the omphalos hypothesis, by the appeal to parsimony? The challenge to the proponents of the argument from design (and similarly for intelligent design) is to make precise testable predictions, not postdictions that distinguish it from natural selection. It is only the success of such predictions that will establish the argument from design as valid. To date, they have not been convincing.

63. THE ILLUSION OF PURPOSE

When something happens, people always want to know why: What is the purpose? Why did hurricane Sandy hit New York? Was it punishment for past sins? The idea of purpose is so central to people's thinking that we want a purpose for every happening. This was engrained in philosophy as Aristotle's final cause. Aristotle regarded the final cause as the most important of his four causes and it became central to medieval philosophy. Understanding the final cause has, indeed, been important to our survival. It was important to know the reason for the lioness taking a stroll. Was she just doing it for exercise or was she looking for a meal? If the latter, it was very important to give her a wide berth. Similarly in social interactions, it is important to know the purpose behind a person's behaviour. Are they just being nice or do they have ulterior motives? And if so, what?

Purpose or Aristotle's final cause is entirely absent from the physical sciences and downplayed in science generally. This leads to an argument against evolution. Evolution by natural selection has no purpose but is a stochastic process with the direction of each step being independent of the previous step and depending only on the local conditions at that moment. Apes did not evolve to form the stock from which humans later arose but rather humans arose as a result of local environmental pressures on the ape. The precise argument against evolution by natural selection is that since natural processes have no purpose, purpose could not have arisen unless there was an outside agency to give purpose. Since purpose is seen, for example in animal and human behaviour, such an outside agency must exist. If evolution produced this purpose, it must have been guided by an external purpose and not be due entirely to natural selection.

This argument is a prime example of proof by lack of imagination. It relies on not having enough imagination to find a method for purpose to arise from natural selection. Hence, the precise argument against evolution can be stated as: I can imagine no way that purpose can arise except by an external agency, therefore evolution must be caused by an external agency. The counter to proof by lack of imagination is the *just so story*. That is a story made up to explain a given occurrence without any evidence of widespread validity. Generally, I regard *just so stories* as uninteresting and certainly not science[1]. To make *the just so story* science, one would have to use it to make test-

---

[1] They can however be entertaining.

able predictions. But as a counter to proof by lack of imagination, that is not necessary. All that is necessary is to provide one plausible counterexample. I will now give *a just so story* to counter the argument in the last paragraph.

So, let's see how the illusion of purpose, if not purpose itself, could arise. Consider some bacteria in a solution with a gradient for food. The bacterium that moves towards more food will on average produce more offspring and therefore the population will eventually be dominated by those that move up the gradient. The resulting behaviour appears to have a purpose: namely to get more food. However, it is just the response to the local conditions, conditioned by evolution's feedback loop.

One can apply the same type of reasoning to more complex situations and in every case evolution favours those individuals whose behavior appears to have purpose. Consider the case of a bird building a nest. Birds that build nests that do a better job of protecting their young will have more offspring (balanced somewhat by the cost of building the nest). Similarly with the behaviour of young men courting young women (and vice versa). Those that are successful produce offspring while those that aren't, don't reproduce. Thus, the behaviour seems to have a purpose but in reality, it is only that those who behave in a certain way leave offspring and hence, that behaviour is all that survives. Incidentally, this may also explain why there are so few geeks.

Thus, we see that purpose, or to be more precise, the illusion of purpose, can arise from the feedback loop in evolution. Evolution favours those behaviours that work towards the end of producing more offspring, which is a purely mechanical process. But saying purpose is an illusion is perhaps going too far. In building models of animal and even plant behaviour, purpose is a useful concept that makes the job easier. Models that include purpose are simpler and make better predictions than those without and even if they didn't, we are human after all, and do enjoy a *just so* story. Purpose, like the nucleon, is an emergent property that arises from the underlying dynamics. So the next time you are pursuing a member of opposite sex with a definite purpose in mind, remember that purpose is, if not an illusion, just an emergent property.

64. THE ROLE OF MATHEMATICS AND RATIONAL ARGUMENTS IN SCIENCE

Mathematics is a tool used by scientists to help them construct models of how the universe works and make precise predictions that can be tested against observation. That is really all there is to it, but I had better add some more or this will be a really short essay.

For an activity to be science, it is neither necessary, nor sufficient, for it to involve math. Astrology uses very precise mathematics to calculate the planetary positions, but that does not make it science any more than using a hammer makes one a carpenter (Ouch, my finger!). Similarly, not using math does not necessarily mean one is not doing science any more than not using a hammer means one is not a carpenter. Carl Linnaeus's (1707 – 1778) classification of living things and Charles Darwin's (1809 – 1882) work on evolution are prime examples of science being done with minimal mathematics (and yes, they are science). The ancient Greek philosophers, either Plato or Aristotle, would have considered the use of math in describing observations as strange and perhaps even pathological. Following their lead, Galileo was criticized for using math to describe motion. Yet since his time, the development of physics, in particular, has been joined at the hip to mathematics.

The foundation of mathematics itself is a whole different can of worms. Is it simply a tautology, with symbols manipulated according to well-defined rules? Or is it synthetic *a priori* information? Is 2+2=4 a profound statement about the universe or simply the definition of 4? Bertrand Russell (1872 – 1970) argued the latter and then showed 3+1=4. Are the mathematical theorems invented or discovered? There are ongoing arguments on the topic, but who knows? I certainly don't. Fortunately, it does not matter for our purposes. All we need to know about mathematics, from the point of view of science, is that it helps us make more precise predictions. It works, so we use it. That's all.

I could end this essay here, but it is still quite short. Luckily, there is more. Mathematics is so entwined with parts of science that is has become its de facto language. That is certainly true of physics where the mathematics is an integral part of our thinking. When two physicists discuss, the equations fly. This is still using mathematics as a tool, but a tool that is fully integrated in to the process of science. This has a serious downside. People who do not have a strong background in mathematics are to some extent alienated from science. They can have, at best, a superficial understanding of it from studying the translation of the mathematics into common language. Something is always lost in a translation. In translating topics like quantum mechanics—or indeed most of modern particle physics—that loss is large; hence nonsense like the *God Particle*. There is no *God Particle* in the mathematics, only some elegant equations and, really, considering their importance, quite simple equations. One hears question like: how do you really understand quantum mechanics? The answer is clear, study the mathematics. That is where the real meat of the topic is and where the understanding is—not in some dreamed up metaphysical interpretation.

Closely related to mathematics are logical and rational arguments. Logic may or may not give rise to mathematics, but for science, all we require from logic is that it be useful. Rational arguments are a different story. Like mathematics, they are useful only to the extent they help us make better predictions. But that is where the resemblance stops. Rational arguments masquerade as logic, but often become rationalizations: seductive, but specious. Unlike mathematics, rational arguments are not sufficiently constrained by their rules to be 100% reliable. Indeed, one can say that the prime problem with much of philosophy is the unreliability of seemingly rational arguments. Philosophers using supposedly rational arguments come to wildly different conclusions: compare Plato, Descartes, Hume, and Kant. This is perhaps the main difference between science and philosophy: philosophers trust rational arguments, while scientists insist they be very tightly constrained by observation; hence the success of science.

In science, we start with an idea and develop it using rational arguments and mathematics. We check it with our colleagues and convince ourselves using entirely rational arguments that it must be correct, absolutely, 100%. Then the experiment is performed. Damn—

another beautiful theory slain by an ugly observation. Philosophy is like science, but without the experiment[1]. Perhaps the real definition of a rational argument, as compared to a rationalization, is one that produces results that agree with observations. Mathematics, logic, and rational arguments are just a means to an end, producing models that allow us to make precise predictions. And in the end, it is only the success of the predictions that count.

65. IS SCIENCE CONSISTENT WITH EVOLUTION?

The evolutionary argument against naturalism.

Alvin Plantinga (b. 1932), professor emeritus of philosophy at the University of Notre Dame, is a leading theistic philosopher and opponent of evolution. He has proposed an intriguing, and specious—yet nonetheless intriguing—argument against evolution. It is intriguing for several reasons: first, because on the face of it, it is plausible. Second, because it is typical of a whole class of specious arguments. Finally, because it highlights the difference between how scientists and philosophers approach a problem.

The argument runs as follows[2]:

> *The naturalist can be reasonably sure that the neurophysiology underlying belief formation is adaptive, but nothing follows about the truth of the beliefs depending on that neurophysiology. In fact, he'd have to hold that it is unlikely, given unguided evolution, that our cognitive faculties are reliable. It's as likely, given unguided evolution, that we live in a sort of dream world as that we actually know something about ourselves and our world.*

In other words, if people evolved, they could not trust their cognitive faculties to give them the truth and hence, do science. He goes on to

---

[1] I believe this observation comes from one of the Huxleys but I cannot find the reference.

[2] From http://www.booksandculture.com/articles/2007/marapr/1.21.html

argue that it is only possible to trust our cognitive faculties if people are created in God's image.

It is amusing that unbelievers argue the opposite; namely that the existence of a God means science is impossible since he/she/it could override the rules of nature at will and there would be no reason to assume constant laws. Both are correct to this extent: Absolute knowledge is impossible,[1] independent of God's existence. But back to Plantinga's argument; it hinges on the concept of truth, or equivalently, reliability. But what is truth? A profound question—or a meaningless one. The difference between profound and meaningless is often vanishingly small.

At one level, the idea of truth is simple: Does the testimony of the person on the witness stand agree with what happened? Or perhaps the simpler question: Does the testimony agree with what the person believes happened? The second is a less stringent requirement. But from this simple concept, *the grand metaphysics concept of* TRUTH™ is generated. Whatever this grand metaphysical concept is, science is not concerned with it. Is it TRUTH™ that colds are caused by viruses? The reductionist, at least if he believes in string theory, would say no. Colds, like all other phenomena, are caused by how strings vibrate in eleven dimensions. Viruses are just a wimpy low-energy approximation to the real TRUTH™.

In science, we build models for how the universe works, which usually have a limited range of validity. Think of classical mechanics which is only valid for velocities much less than the speed of light. Is classical mechanics the TRUTH™? No, certainly not, it fails in various places. But it is certainly useful. Science is a natural extension of the model building the unconscious mind does all the time, model building that is necessary for us to survive in a hostile world. The surprising thing is not that beings who evolved created science, but rather, that they did not do it sooner. Plantinga's problem is that he does not understand what science is or how it works—seeking effective models rather than the TRUTH™, whatever that may be. As a philosopher, he should have known better than to try to falsify evolu-

---

[1] See the essay *Let the Mystery Be.*

tion, since by the Duhem-Quine thesis, no model can be falsified. Arguing that the current models have deficiencies is never enough. You have to provide better ones with more predictive power.

In the same manner that Plantinga's argument relies on the grand metaphysics concept of TRUTH™, many arguments in philosophy rely on similar word definitions. A prime example is the ontological agreement for God's existence. First proposed by Anselm of Canterbury (1033 – 1109), the argument goes as follows (from WIKIPEDIA):

> *Define God as the greatest possible being we can conceive.*
> *If the greatest possible being exists in the mind, it must also*
> *exist in reality. If it only exists in the mind, a greater being is*
> *possible—one which exists in the mind and in reality.*

Note that his argument hinges on the definition of greatest. My daughter believes that anything, no matter how great, can be made greater by being pink. Thus the greatest being is pink. If I define nonexistence as being greater than existence,[1] the ontological argument becomes an argument for God's nonexistence. Evil is another word that is frequently made into a *grand metaphysical concept*, EVIL™, and used to justify various philosophical positions. The concept of actions I do not like is then even taken a step further and personified in the concept of the devil.

While our concepts and word definitions may reflect reality, they do not constrain it. Experience is determined only by experience. In the end, models founded on observation take precedence over philosophical arguments based on word definitions and phenomenologically unconstrained speculations. If such philosophical arguments disagree with scientific models, so much the worse for them. Thor showing up for Thursday afternoon tea at the Empress Hotel would make all arguments regarding his existence moot[2]. One observation is worth a thousand philosophical arguments.

---

[1] See Ecclesiastes 4:1–4 or Arthur Schopenhauer's (1788 – 1860) philosophy for why this definition may be reasonable.
[2] You can tell it is Thor because he would be carrying a large hammer and one of the goats pulling his chariot would be limping.

Jargon, even the name has a harsh ring to it. Can anyone but an author love a title like[1]: WALKING NEAR A CONFORMAL FIXED POINT: THE 2-D O(3) MODEL AT THETA NEAR PI AS A TEST CASE? Ugh! *How can anyone take science seriously when it uses so much jargon?* said the teamster[2] as he told his helper to fasten the traces to the whiffletree and check the tugs and hames' straps. Jargon is everywhere and not unique to science. While you may not understand what the teamster is talking about, my father would have understood instantly and then gone to get a jag of wood.

But back to jargon. To the uninitiated the above title, like the teamster's words, seems like so much gobbledygook. But to the initiated, those working in the field, it is a precise statement and easily understood. Trying to put the title, or the teamster's words, in a form understandable to the layperson would have been a fool's errand. In making it understandable to a more general audience, the precision would have been lost and we would probably never have gotten that jag of wood. That would have been unfortunate as Nova Scotian winters can be cold.

One of the principles of all good writing is to tailor the communication to the intended audience. When I am helping put together a report for TRIUMF, the instructions to the authors always includes a statement about the intended audience. Even then, the good authors frequently ask me to make the description of the intended audience more precise. Life gets more complicated when a document has more than one intended audience. Then it is necessary to have a layered document where introductory sections are understandable by an intelligent layperson while the later sections are directed at the specialist. One is reminded of the old joke about the structure of a good seminar: The speaker starts at a low level understandable by anyone and then as the seminar progresses he becomes more technical and less understandable so that by the end, even the speaker does not know

---

[1] First title on the lattice archive the day I checked to get an example.
[2] The kind that drives horses.

what he is talking about. Well, perhaps that is getting a little too carried away, but one can err on either side, by making the writing too technical for the audience or not technical enough.

Similarly, the reader has to realize that the writing may not be directed at him or her. We, as people with technical expertise, have to be careful not to judge non-technical writing too harshly because it does not capture all the subtle nuances we are aware of. Including them would lose the layperson. It is a fine line between not confusing the layman and misleading him. When I am reading an article directed at a general audience, on a topic I am an expert in, I find I have to translate the layman's language back to the technical language before I can understand it. That is as it should be.

Conversely, in fields we are not experts in, we should not criticize technical writing as being too filled with jargon. This latter mistake is made frequently by politicians and commentators who criticize technical writing due to ignorance. Few have the wisdom of the former Canadian Prime Minister, Pierre Elliot Trudeau, who said on opening TRIUMF, *I do not know what a cyclotron is, but I am glad Canada has one.* It is a rare politician who has the confidence to admit ignorance. As an undergraduate student, I picked up a copy of M.E. Rose's (1911 – 1968) book: ELEMENTARY THEORY OF ANGULAR MOMENTUM (published 1957). That is when I learned one should be leery of books with *elementary* in the title[1]. If that is an elementary book, I would hate to have to read an advanced one. It is a good book, but I, at that stage in my career, was not the intended audience.

Words only have meaning within the context they are used. When used with a person possessing a similar background, the context does not have to be spelled out. Thus, in conversation with a colleague I have worked with for some time a lot is understood without being stated explicitly. Jargon speeds up communication and makes it less prone to misunderstanding. On the other hand, with people who are not acquainted with the field, we have to spell out the background

---

[1] Books with *elementary* in the title are usually advanced while those with *advanced* in the title are usually elementary.

assumptions and suppress the details that are only of interest to the expert.

In the end, it is quite unfortunate that jargon has been abused and hence has received a bad name. In technical writing, jargon or technical terms are not only acceptable but necessary. So press on and employ jargon—but only where appropriate.

67. SCIENCE AND ENGINEERING: *VIVE LA DIFFÉRENCE*

This essay was motivated by a question from an engineering colleague. It would be presumptuous to say *friend*, as scientist and engineers are in a state of *friendly* rivalry, however, not to the extent of rivalry between science and arts. I once saw a sign in a university engineering department hallway that read: *Friends do not let friends study arts[1]*. Be that as it may, my colleague's question was why scientists do not show the same order in all their work as they show in writing papers. This question I will attempt to answer in this essay.

Engineering is far older than science, being perhaps the second oldest profession, dating back past the building of the pyramids (Imhotep from the 27th century BCE is the oldest named engineer), to Stonehenge, and probably back to when the first club was engineered. Stonehenge is amazing as it was probably built without the documentation that is the hallmark of modern engineering practice. Unfortunately, that means we do not know what the objectives were and this has led to much futile speculation as to its purpose.

Science and engineering are sibling disciplines, frequently mentioned together and have much in common. The main similarity is that they both deal with the observable universe and are judged by their ability to make correct predictions regarding its behaviour. For example, that the Higgs boson would be found at the Large Hadron Collider (LHC) or that the building will not collapse in an earthquake. Secondarily they use similar techniques, placing high importance on analytic rea-

---

[1] Since my wife has a B.A. in English, I do not agree with this.

soning, to the extent that Asperger's syndrome is sometimes called the engineer's disease. The relation between Asperger's syndrome and engineers or scientists may be an urban myth but it does indicate the relation of extreme analytic thought to both science and engineering. The solution to problems in both relies on the same problem solving skills, creativity, analytic thinking and mathematics. Do not let anyone tell you that either does not require a high degree of intellectual activity.

Science and engineering rely on each other. Behind every engineering project is a great deal of science, from the basic understanding of Newtonian mechanics in the building of a bridge to the advanced materials science in the construction of a cell phone. Actually, the cell phone is a good example of all the science needed: it depends on Newtonian mechanics (the construction of the cell phone towers), quantum mechanics (the operation of the transistors), classical electromagnetism i.e. Maxwell's equations (the propagation of the signal from the tower to the cell phone), materials science (almost all the cell phone itself), and general and special relativity (the GPS timing that is necessary in some cell phone technologies).

Equally, science is beholden to engineering. From simple things like the buildings that house scientific equipment to complicated things like the ATLAS detector at the LHC. Making a building may seem simple but, as I see with the new ARIEL building at TRIUMF, nothing is simple and even something as basic as a laboratory building relies on engineering expertise. The ATLAS detector is another story. Its size and complexity are a marvel of engineering virtuosity. Back to TRIUMF, the IEEE [1] has recognized the TRIUMF cyclotron, commissioned in 1974 and the main driver for much of TRIUMF's science program, as an Engineering Milestone. Even the slide rule I used back in ancient history as an undergraduate [2] was an engineering achievement.

---

[1] Institute of Electrical and Electronics Engineers.
[2] HP produced the first pocket calculator when I was an undergraduate student.

Despite the close relationship between science and engineering the two are different. The difference can be summarized in this statement: *In engineering you do not start a project unless you know the answer while in science you do not start a project if you know the answer.* Engineering is based on everything being predictable; you do not start building a bridge unless you know you can complete it. In science, the purpose of a project is to answer a question to which the answer is currently unknown. For example, if the properties of the Higgs boson were known, it would not have been necessary to build the LHC. Good engineering practice is based on order but at the center of science is chaos. We are exploring the unknown; great discoveries can come from serendipity. In science, something not working as expected can lead to the next big breakthrough. In engineering, something not working as expected can lead to the bridge collapsing. Advances in science are frequently due to creativity, not to following rules.

This difference in perspective leads to very different cultures in the two disciplines. The engineer is much more concerned with process and following procedure; the scientist with following up his most recent hunch—after all, it could lead to a Nobel Prize. Engineering versus science: order versus creative chaos. This is clearly an oversimplification as there is no clean separation between engineering and science[1] and large science is making scientists more like engineers, but it is a good indication of the divergence between the two mindsets. Thus, although engineering and science are closely related and indeed intertwined, the two, in their heart of hearts, are very different; engineering uses science in order to build and science uses engineering in order to explore.

69. MEASUREMENT AND THE NEW SI UNITS

The Système international d'unités (SI) unit definitions[2] will be changing again in the next few years (currently scheduled for 2018). You would think that choosing the units of measurement would be an

---

[1] R&D (research and development) can fall somewhere in between the two.
[2] Colloquially known as the metric system.

unemotional topic, but as I recall from Canada's only partially successful attempt to convert to the metric system, that is far from the case. I remember one rather irrational editorial on the topic where the writer went on about how the changing definition of the metre was an indication that the people behind the metric system did not know what they were doing. Since this was in an English Canadian paper, he blamed the problem on the French for having blown the original definition. **Ignorance profound**. The writer would probably have been surprised to learn that the inch is defined as 2.54 centimetres except, of course, in the US where there is a second inch (the surveyor's inch) defined as 39.37 inches equal one metre. Ah, the joy of traditional measurements. There are at least three different gallons in use, and as for barrels, there are more than you can shake a stick at. However, the petroleum barrel is defined as exactly 158.987294928 litres. I am sure you wanted to know that and don't forget the last decimal place—the 8 is very important. As far as I can see, the only reason for using the traditional units is familiarity and yes, I still use the inch and foot, but also the kilometre. And I believe it's also safe to say, that the generation born after the country officially switched, also does the same. That is the joy of living in a country that has partially converted to metric.

Measurements tend to be of two types. One is pure numbers like the number of ducks in a row (or in a pond). The other type is the measurement of a number with a dimension. Here we need a standard to compare against; a length of six feet only makes sense if we know what a foot is. In other words, we have a standard for it. Thus, the need to define units so different people can compare their results, and when we buy a hogshead of beer, we know how much we are getting.

Editorial writers will have another chance to rant in a few years as the General Conference on Weights and Measures is set to change the definitions of the basic metric or SI units again—this time, not the metre but the kilogram and other units. The history of how the definitions of the units have changed over time is quite interesting, involving not just changing technology but also changing tastes. The original metre was defined in terms of the distance from the equator to the North Pole. But this could not be determined sufficiently accurately, so the standard was shifted to a physical artifact; a rod kept in Paris with two marks on it. This was then shifted to the wavelength of light from a certain atomic transition and finally, to fixing the speed of light. Similarly, for time, the second went from being defined in terms of the length of the day to being defined in terms of the frequency of an atomic transition. There is a trend from defining the units in terms of macroscopic quantities—the size of the earth, the length of day, the length of a bar—to microscopic quantities, or more specifically, atomic properties. There is a simple reason for this, namely that it is in atomic systems that the most accurate measurements can be made. Unfortunately, it also makes the unit definitions esoteric and detached from everyday experience. Everyone can identify with the length of a foot, but it is not immediately clear what the speed of light has to do with distance. Telling my daughter it takes three nanoseconds for light to travel from her head to her foot doesn't do much for her. There is also a trend, partly aesthetic, towards defining the base units by fixing the fundamental constants of nature.

A fundamental constant of nature, like the speed of light, starts its life as something that relates two apparently unrelated quantities. In the case of the speed of light, it is time and distance. But then over time, it comes to be just a way of relating different units for measuring the same thing. Indeed, time units are sometimes used for distances and vice versa. This even happens in everyday life, such as when the distance from Vancouver to Seattle is given as three hours, meaning, of course, an average travel time. But in science, the relation is more definite and defining the metre in terms of the speed of light makes it explicit that the fundamental constant, the speed of light, is just a conversion factor from one set of units to another, from seconds to metres (1 metre = 3.3 nanoseconds).

The new proposal for the base SI units continues this trend of defining units by fixing fundamental constants. The degree Celsius is now defined in terms of the properties of water—the so called triple point. In the proposed new system, it will be defined by fixing a fundamental constant, the Boltzmann constant. The Boltzmann constant relates degrees to energy. At the microscopic level, i.e. in statistical mechanics, temperature is just a measure of energy and the new definition of the degree makes this explicit. Again, a fundamental constant turned to a conversion factor between different units—degrees and joules.

179

The case of the kilogram is more subtle. It is currently defined by a physical artifact—the standard kilogram stored in Paris. The new proposal is to determine the kilogram by fixing the fundamental constant; Planck's constant. This is another example of a fundamental unit becoming just a conversion factor between different units, in this case between time and energy units, or equivalently distance and momentum units.

As a theorist, this new set of units makes it nice for me as I like to use what are called natural units in my calculations. These are given by setting the speed of light (c), Planck's constant (ħ), Boltzmann's[1] constant (k) and π all equal to 1 (Okay, usually not π, but I did see that legitimately done once). An interesting side effect of the new units is that they all have exact conversion from these natural units. There is another set of natural units called Planck units which are defined in terms of the gravitational strength and the strength of the electromagnetic force. (In the proposed change, the charge of the electron is used to define the electromagnetic units.) Ultimately, those may be the most elegant units but we are nowhere close to having the technology to make them the basis of the SI units.

Naturally, any change of units has the naysayers coming out of the woods. One of the criticisms of the new units is that, since the fundamental constants are fixed by definition, we can no longer study their time dependence. To some extent, this is true. For example, with the current definition of the kilogram, Planck's constant changes every time atoms are lost or gained by the standard kilogram. This change will be lost with the new units. This illustrates the absurdity of asking if a fundamental constant changes in isolation. All that is meaningful is if the constant has changed with respect to some other quantity with the same dimensions. The new choice of units makes this explicit, which is a good thing.

---

[1] Named after Ludwig Boltzmann (1844 – 1906).

69. SCIENCE: THE ART OF THE APPROPRIATE APPROXIMATION

There is this myth that science is exact. It is captured nicely in this quote from an old detective story (Jacques Futrelle (1875 – 1912), THE SILVER BOX, 1907):

> *In the sciences we must be exact—not approximately so, but absolutely so. We must know. It isn't like carpentry. A carpenter may make a trivial mistake in a joint, and it will not weaken his house; but if the scientist makes one mistake the whole structure tumbles down. We must know. Knowledge is progress. We gain knowledge through observation and logic—inevitable logic. And logic tells us that while two and two make four, it is not only sometimes but all the time.*

Unless, of course, it is two litres of water and two litres of alcohol, then we get less than four litres. Note also the almost quaint idea that science is certain, not only exact, but certain. *We must know.* The view expressed in this quote is unfortunately not confined to century-old detective stories, but is part of the modern mythology of science. But in reality, science is much more like carpentry. A trivial mistake does not cause the whole to collapse, but I would not like to live in a house built by that man.

To the best of my knowledge, there has never been an exact calculation in all of physics. In principle, everything in the universe is connected. The earth and everything in it is connected by the gravitational field to the distant quasars. But you say, surely that is negligible, which is precisely the point. It is certainly not exactly zero, but with equal certainty, it is not large enough to be usefully included in any calculation. I know of no terrestrial calculation that includes it. Even closer objects like Jupiter have negligible effect. In the grand scheme, the planets are too far from the earth to have any earthly effect. Actually, it is not the gravitational field itself which is important but the tidal forces which are down an additional factor of the ratio of the radius of the earth to the distance to the planet in question. Hence, one does not expect astrology to be valid. The art of the appropriate approximation tells us so.

Everywhere we turn in science we see the need to make the appropriate approximations. Consider numerical calculations. Unless you are calculating the hypotenuse of a triangle with side of 3 and 4 units, almost any numerical calculation will involve approximations. Irrational numbers are replaced with rational approximations, derivatives are replaced with finite differences, integrals with sums, and infinite sums with finite sums. Every one of these is an approximation—usually a valid approximation—but nevertheless an approximation. Mathematical constants are replaced by approximate values. Someone once asked me for assistance in debugging a computer program. I noticed that he had pi approximated to only about six digits. I suggested he put it in to fifteen digits (single precision on a CDC computer). That, amazingly enough, fixed the problem. Approximations, even seemingly harmless ones, can bite you.

Even before we start programing and deciding on numerical techniques, it is necessary to make approximations. What effects are important and which can be neglected? Is the four-body force necessary in your nuclear many-body calculation? What about the five-body force? Can we approximate the problem using classical mechanics, or is a full quantum treatment necessary? Thomas Kuhn (1922 – 1996) claimed that classical mechanics is not a valid approximation to relativity because the concept of mass is different. Fortunately, computers do not worry about such details and computationally classical mechanics is frequently a good approximation to relativity. The calculation of the precision of the perihelion of Mercury does not require the full machinery of general relativity, but only the much simpler post-Newtonian limit. And on and on it goes, seeking the appropriate approximation.

Sometimes the whole problem is in finding the appropriate approximation. If we assume nuclear physics can be derived from quantum chromodynamics (QCD), then nuclear physics is reduced to finding the appropriate approximation to the full QCD calculation, which is by no means a simple task. Do we use an approximation to the nuclear force based on power counting, or the old fashioned unitarity and crossing symmetry? (Don't worry if you do not know what the words mean, they are just jargon and the only important thing is that the approximations lead to very different looking potentials.) Do the results depend on which approach is used, or only the amount work required to get the answer?

Similarly, in materials science, all the work is in identifying the appropriate approximation. The underlying forces are known: electricity and magnetism. The masses and charges of the particles (electrons and atomic nuclei) are known. It *only* remains to work out the consequences. *Only*, he says, *only*. Even in string theory, the current proposed theory of everything, the big question is how to find useful approximations to calculate observables. If that could be done, string theory would be in good shape. Most of science is the art of finding the appropriate approximation. Science may be precise, but it is not exact, and it is in finding the appropriate approximation that we take delight.

70. IN DEFENSE OF MICKEY MOUSE SCIENCE

Give it to me—the real news

So I will

Well, Dadamashay, let me see what skill you have. Tell me the big new news of these days, making it ever so small.

Listen[1]

When, I was a graduate student, somewhat after the time of the Vikings in long boats, my thesis supervisor, Prof. Rajat K. Bhaduri[2], took me with him when he went on sabbatical to Copenhagen, a Mecca for nuclear physics at that time. When we were leaving there, his officemate gave him a small Mickey Mouse figurine so he would know what kind of physics to work on. *Well another man might have been angry, And another man might have been hurt, But another man never would have*[3] stressed during his seminar that he was using a

---

[1] Quoted from Rabindra Nath Tagore (1861 – 1941) in *Fables*. Also used as an inscription in R.K. Bhaduri's book: MODELS OF THE NUCLEON.
[2] A scholar and a gentleman.
[3] With apologies to Harry Chapin and the song: THE TAXI.

Mickey Mouse model. Ah yes, Mickey Mouse science, the simple model or calculation that brings out salient features that are all too often lost or obscured in the complete calculation.

We all know what big science is: the big detectors at the Large Hadron Collider (CMS has a 12,500 ton steel yoke) or the Super-Kamiokande (50,000 tons of water). That is big science. Even theoretical physics does big science: the massive calculations of lattice quantum chromodynamics (QCD) or the nuclear shell model. Now, there have been attacks on big science, either the LHC or lattice QCD, as being inherently evil because they are so big. Would you believe, even books written on the topic? I strongly disagree with that view. Large science is an essential part of science. Big is needed to answer the questions we want answered. However, there is more to science than that. We need the little to complement the big, the simple to complement the complex. As a post-doc, I was returning from a somewhat annoying conference with Gerry Brown[1] (1926 – 2013), one of leading nuclear physicists of that generation, when he turned to me in exasperation and said that people did not realize how many hours of computer time went into his simple estimates. There is an interesting concept: using computer time to justify simple estimates, simple complementing the complex. The purpose of computing is insight, not numbers[2] and the simple Mickey Mouse models are essential in generating that insight—even when they are justified by complex calculations.

The simple models are useful in a number of ways. First, they are useful in checking the results of complex computer calculations. I have learnt through bitter experience never to believe the result of a computer calculation until I have *understood* them (and not always then). That is, until using some simple model or estimates, either explicitly or implicitly, I can reproduce the main trends of the results. In trying to do that, I have frequently found errors. Never trust a number you do not understand.

---

[1] No, not the California politician.
[2] Quoted from Richard Hamming (1915 – 1998).

184

Second, we want to understand which aspects of the model are important in reproducing the results and which are coincidental. Scientific models are designed to predict future observations, but which aspects of the model are crucial to that endeavor? It is through the use of simple models that we can most easily explore the dependencies of the results on the assumptions. We calculate some nuclear cross-section. Is that bump significant? What, if anything, does the location of the bump tell us? What about the turn up near threshold? Is that an artifact? We want to know more than merely if the calculation fits the data. It is here that the simple models come in. They give us the insight into how the models can be improved and what assumptions are not necessary and can be eliminated.

Finally, and most importantly, it is the simple models that allow us, as people, to understand the results. It is not just for the layman that we need the simple models, but for the expert as well. A prime example would be the non-relativistic quark model. Its success calculating the properties of the excited states of the proton was touted as proof of the quark model but all it tested was the symmetries built into the calculations. The simple approximations to the non-relativistic quark model revealed its pretentions. But as a Mickey Mouse model, the non-relativistic quark model gave us insight into QCD that would have been difficult if not impossible to obtain otherwise.

I suppose one could hook up the computers directly to the experiments and have them generate models, test the models against new observations and then modify the experimental apparatus without any human intervention. However, I am not sure that would be science. Science is ultimately a human activity and the models we produce are products of the human mind. It is not enough that the computer knows the answer. We want to have some feeling for the results, to understand them. Without the simple models, Mickey Mouse science, that would not be possible: the big news made ever so small.

71. CAUSE AND EFFECT: A CORNERSTONE OF SCIENCE OR A MYTH?

Cause and effect has been central to many arguments in science, philosophy and theology down through the ages, from Aristotle's four causes[1] down to the present time. It has frequently been used in philosophy and Christian apologetics in the form: *The law of cause and effect is one of the most fundamental in all of science.* But it has its naysayers as well. For example, Bertrand Russell (1872 –1970)[2]:

> *All philosophers, of every school, imagine that causation is one of the fundamental axioms or postulates of science, yet, oddly enough, in advanced sciences such as gravitational astronomy, the word "cause" never occurs. ... The law of causality, I believe, like much that passes muster among philosophers, is a relic of a bygone age, surviving, like the monarchy, only because it is erroneously supposed to do no harm.*

You can accuse Russell of many things, but being mealy-mouthed is not one of them. Karl Pearson (1856 – 1936), who has been credited with inventing mathematical statistics, would have agreed with Russell. He never talked about causation, though, only correlation.

One of the people who helped elevate cause and effect to its exalted heights was David Hume (1711 – 1776). He was a leading philosopher of his day and known as one of the British empiricists (in contradistinction to the continental rationalists). Hume was one of the first to realize that the developing sciences had undermined Aristotle's ideas on cause and effect and he proposed an alternate in two parts: first, Hume defined cause as *an object, followed by another, and where all objects similar to the first are followed by objects similar to the second.* This accounts for the external impressions. His second definition, which defines a cause as *an object followed by another, and whose appearance always conveys the thought to that other,* captures the internal sensation involved. Hume believed both were

---

[1] Discussed in a previous essay.
[2] From the essay: ON THE NOTION OF CAUSE.

needed. In thus trying to relate cause and effect directly to observations, Hume started the philosophy of science down two dead end streets: one was the idea that cause and effect was central to science and the other led to logical positivism.

Hume's definitions are seriously flawed. Consider night and day. Day invariably follows night and the two are thought of together but night does not cause day in any sense of the word. Rather, both day and night are caused by the rotation of the earth, or, if you prefer, a geocentric frame, by the sun circling the earth. The true cause has no aspect of one thing following another or one causing thought of the other. And the cause does not have to in any way resemble the effect. One can find many other similar cases: the increasing light of dawn does not cause the sun to rise despite the increasing light before the sun rises; it is the sun rising that causes the light of dawn. Trees losing their leaves does not cause winter but rather the days getting shorter causes the trees to lose their leaves and is a harbinger of winter; the root cause being the tilt of the earth's axis of rotation with respect to the ecliptic.

As just seen, cause and effect is much more complicated than Hume and his successor thought, but not nonexistent as its detractors maintain. In the words of the statistician: correlation does not imply causation. However, it can give a strong hint. The cock crowing does not cause the sun to rise but the correlation does suggest that the sun rising might just motivate, if not cause, the cock to crow. Similarly, consider lung cancer and smoking. Not all people who smoke get lung cancer and not all people who get lung cancer smoke (or inhale second hand smoke). Nevertheless, there is a correlation. It was this correlation that started people looking to see if there was a cause and effect relation. Here we have correlation giving a hint; a hint that needed to be followed up. And it was followed up. Nicotine was found to be carcinogenic and the case was made convincing. A currently controversial topic is global warming and human activities. Here, as with smoking causing cancer, we have both correlation and a mechanism (the greenhouse effect of carbon dioxide and methane).

Cause and effect went out of favor as a cornerstone of science about the time quantum mechanics was developed. Quantum mechanics is non-deterministic with events occurring randomly. Within the context of quantum mechanics, there is no reason or cause for an atom to decay at one time and not at another. The rise of quantum mechanics and the decline in the prominence of cause and effect are probably indeed cause and effect[1]. However, even outside quantum mechanics there are problems with cause and effect. Much of physics, as Russell observed, does not explicitly use cause and effect. The equations work equally well forwards or backwards, deriving the past from present as much as the future from the past. Indeed, the equations of physics can even propagate spatially sideways rather than temporally forwards or backwards.

In spite of all that, the idea of cause and effect is useful. To understand its limitations and successes we have to go back to one of my mantras: the meaning is in the model. Cause and effect is not something that can be immediately deduced from observation, as Hume implies, but it is not a meaningless concept as Russell said or the physics discussion above might seem to imply. Rather, when we develop our models for a particular situation the idea of causation comes out of that model, is part and parcel of the model. We believe that the post causes the shadow and not the other way around, because of our model on the nature of light and vision. Similarly, the idea that the earth's rotation causes day and night comes out of our model for light, vision and the solar system. The first chapter of Genesis indicates that this was not always considered obvious[2]. That smoking causing lung cancer is part of the biological model for cancer. That human activities causing global warming comes out of atmospheric modeling. But arising from a model does not make cause and effect any less real nor the concept less useful. Identifying smoking as a cause of cancer has saved human lives and identifying carbon dioxide and methane as the main causes of global warming will, hopefully, help save the world. Cause and effect may not a cornerstone of science but it is still a useful concept and certainly not *a relic of a bygone age.*

---

[1] Quantum mechanics also wreaks havoc with the related philosophical principle of sufficient reason.
[2] Day and night were created before the sun.

72. SIMPLICITY: THE SECRET SAUCE IN THE SCIENTIFIC METHOD

Simplicity plays a crucial, but frequently overlooked, role in the scientific method. Considering how complicated science can be, simplicity may seem to be far from a driving source in science. Is string theory really simple? If scientists need at least six, seven or more years of education and training past high school, how can we consider science to be anything but antithetical to simplicity?

Good questions, but simple is relative. Consider the standard model of particle physics. First, it is widely agreed upon what the standard model is. Second, there are many alternatives to the standard model that agree with the Standard Model where there is experimental data but disagree elsewhere. One can name many[1]: Little Higgs, Technicolor, Grand Unified Models (in many varieties), and Super Symmetric Grand Unified Models (also in many varieties). I have even attended a seminar where the speaker gave a general technique to generate extensions to the standard model that also have a dark matter candidate. So why do we prefer the standard model? It is not elegance. Very few people consider the standard model more elegant than its competitors. Indeed, elegance is one of the main motivations driving the generation of alternate models. The competitors also keep all the phenomenological success of the standard model. So, to repeat the question, why do we prefer the standard model to the competitors? Simplicity and only simplicity. All the pretenders have additional assumptions or ingredients that are not required by the current experimental data. At some point they may be required as more data is made available but not now. Thus we go with the simplest model that describes the data.

This is true across all disciplines and over time. The elliptic orbits of Kepler (1571 – 1630) were simpler than the epicycles of Ptolemy (c. 90 – c. 168) and the epicyclets of Copernicus (1473 – 1543). We draw straight lines through the data rather than 29th order polynomials. If the data has bumps and wiggles, we frequently assume they are ex-

---

[1] This list is time dependent and may be out of date.

perimental error. Theoretical curves rarely go through all the data points and no one would take them seriously if they fit every single bump and wiggle. Simplicity is more important than religiously fitting each data point.

Going from the sublime to the ridiculous consider Russell's teapot. Bertrand Russell (1872 – 1970) argued as follows[1]:

> *If I were to suggest that between the Earth and Mars there is a china teapot revolving about the sun in an elliptical orbit, nobody would be able to disprove my assertion provided I were careful to add that the teapot is too small to be revealed even by our most powerful telescopes. But if I were to go on to say that, since my assertion cannot be disproved, it is intolerable presumption on the part of human reason to doubt it, I should rightly be thought to be talking nonsense.*

But what feature of the scientific method rules out the orbiting teapot? Or invisible pink unicorns? Or anyone of a thousand different mythical beings? Not observation! But they fail the simplicity test. Like the various extensions to the standard model, they are discounted because there are extra assumptions that are not required by the observational data. This is otherwise known as Occam's razor.

The argument for simplicity is rather straightforward. Models are judged by their ability to describe past observations and make correct predictions for future ones. As a matter of practical consideration, one should drop all features of a model that are not conducive to that end. While the next batch of data may force one to a more complicated model, there is no way to judge in advance which direction the complication will take. Hence we have all the extensions of the standard model waiting in the wings to see which, if any, the next batch of data will prefer—or rule out.

The crucial role of simplicity in choosing one model from among the many solves one of the enduring problems in the philosophy of sci-

---

[1] From IS THERE A GOD (1952)

ence. Consider the following quote from Imre Lakatos (1922 – 1974) a leading philosopher of science from the last century[1]:

> *But, as many skeptics pointed out, rival theories are always indefinitely many and therefore the **proving** power of experiment vanishes. One cannot learn from experience about the truth of any scientific theory, only at best about its falsehood:* **confirming instances have no epistemic value whatsoever** [emphasis in the original].

Note the premise of the argument: *rival theories are always indefinitely many.* While rival theories may be indefinitely many, one or at most a very few are always chosen by the criteria of simplicity. We have the one standard model of particle physics not an infinite many and his argument fails at the first step. Confirming instances, like finding the Higgs boson, do have epistemic value.

73. SCIENCE AND EXPLORING THE PAST

Many years ago when I was in a grade-eight math class, I was sitting looking out the windows at the dinosaurs playing. Okay, despite what my daughter thinks, I am not quite that old. What I was looking at was planes (not pterosaurs) circling around in the distance. It turned out that a plane had crashed nearby. It was a Handley Page HPR-7 Herald 211 operated by Eastern Provincial Airlines and all eight people on board were killed. Now, it is sometimes claimed that science cannot explain the past. It's even argued that historical sciences like paleontology, archeology, and cosmology somehow use different methods of discovering the past, than say, determining the reason of a plane crash and that is again different from the method for discovering the laws of nature. In reality, the methods are all the same.

I suppose, in response to the plane crash, people could have sat around and made predictions for future plane crashes but instead they used science to try to discover the past—what had caused the plane to

---

[1] From THE ROLE OF CRUCIAL EXPERIMENTS IN SCIENCE, STUD. HIST. PHIL. SCI. 4 (1974).

crash. In this case it turned out to not be so difficult. The Aviation Safety Network describes the cause thus: *Failure of corroded skin area along the bottom centre line of the aircraft beneath stringer No.32 which resulted in structural failure of the fuselage and aerial disintegration.* This was found out by a metallographic examination which provided clear evidence of stress corrosion in the aluminum alloy, although there were some claims that the corrosion was caused by spilled soup seeping through the galley floor. The planes of this type that were remaining in service were repaired to prevent them from crashing as well.

The approach to understanding why the Eastern Provincial Airline's plane had crashed followed a similar approach to any other plane crash: analyze the debris, gather records from the black box and whatever other information is available, and construct a model for what happened. Test the model by making predictions for future observations; for example, that corrosion will be found on other planes of the same type. This sounds very much like the standard scientific method as proposed originally by Roger Bacon (1220 – 1292) and followed by scientists ever since: observe, hypothesize, test, rehypothesize, and repeat as necessary.

The same technique is used for any reconstruction of the past, be it plane crashes, the cause of Napoleon's death, archeology, paleontology, evolution, and cosmology. The cause of Napoleon's death is quite interesting as an exercise in forensic science. The original cause of death was suggested to be gastric cancer. But that is too mundane a cause of death for such an august figure. So the conspiracy advocates went to work and suggested he was poisoned by arsenic. How to test? Easy, look for arsenic in samples of his hair. Well, that was done and arsenic was found. Case closed? Not quite. Were there other sources of arsenic than deliberate poisoning? Yes, the wallpaper in his room had arsenic in it. Also further investigation revealed that he had been exposed to arsenic long before he went to St. Helena. In support of the cancer hypothesis his father also died of stomach cancer. The current consensus is that the original diagnosis was correct. He died of stomach cancer. But notice the play of events: hypothesis—arsenic poisoning, testing—look for arsenic in hair samples, refine hypothesis—check for other sources of arsenic, etc. We can see here the classic process of science being played out in reconstructing the past.

We can continue this technique into the more distant past: When did humans evolve? Why did the dinosaurs die out? How did the earth form? How did the solar system form? What if anything preceded the big bang? All of these questions can be tackled using the standard methods of science. Observations of present tell us about the past, counting tree ring tells us when the tree started to grow.

The interplay between what might be called natural history and natural laws is very intricate. We must interpret the past in order to extract the natural laws and use the natural laws to interpret the past. All our models of science have, explicitly or implicitly, both an historical and a law component. In testing a model for how the universe works—i.e. to develop the laws—we conduct an experiment. Once the experiment is finished, it becomes history and interpreting it is historical science. For example, why did the OPERA experiment claim to see faster than light propagation for neutrinos? Or is the bump seen in searches for the Higgs boson real or an artifact of the detector? Those investigations are as much forensic science as trying to decide why Napoleon died or the dinosaurs became extinct. Thus, all science is historical and sometimes, quite explicitly. Einstein abandoned the cosmological constant based on an alternate model for the history of the universe, namely that it is expanding rather than static.

So, we have science as a unified whole, encompassing the past, present, and future; the natural laws entangled with the natural history. But what about the dinosaurs I did not see out of the math room windows? We can be quite sure they did not exist at that time and that Fred Flintstone did not have one as a pet (a saber-toothed pussy cat is another story). The study of evolution is much like that for plane crashes. You study the debris, in the case of evolution that *debris* includes fossils and the current distribution of species. Consider the fossil *Tiktaalik roseae*, a tetrapod-like fish or a fish-like tetrapod, that was found a few years ago. One can engage in futile semantic arguments about whether it is a fish, or a tetrapod, or a missing link, or whether it is the work of the devil. However, the significant point is that a striking prediction has been confirmed by a peer-reviewed observation. Using evolution, a model of fossil formation, and a model of the earth's geology, a prediction was made that a certain type of fossil would be found in a certain type of rock. *Tiktaalik roseae* dramatically fulfilled that prediction and provides information on the fish-tetrapod transition.

The cause of plane crashes, Napoleon's death, evolution, and the extinction of dinosaurs can all be explored by using the same empirically-constrained model-building techniques as the rest of science. There is only one scientific method.

74. THE MYTH OF THE OPEN MIND

The race of truly open-minded people is long extinct: *To be open minded, I will suspend belief that that tawny blob over there is a leopard.* Pounce, chomp, chomp. Even today, natural selection is working to remove the truly open minded from the gene pool: *To be opened minded, I will remove any judgment of whether jaywalking and texting at the same time is a good or bad idea.* Splat, crumple, crumple. As I said, the race of truly opened-minded people is long extinct, if it ever actually existed.

You may complain that I am misrepresenting the concept of open-mindedness. That is probably true. When most people accuse someone of being closed-minded, they mean little more than that the person does not agree with them. Be that as it may, in general, the related concepts of open-mindedness and freedom from preconceived ideas are vastly overrated. But what about in science? Surely in science it is necessary to keep an open mind and eliminate preconceived ideas? Perhaps, but here is what Henri Poincaré (1854 – 1912) said on the topic[1]:

> It is often said that experiments should be made without preconceived ideas. That is impossible. Not only would it make every experiment fruitless, but even if we wished to do so, it

---

[1] From SCIENCE AND HYPOTHESIS.

could not be done. Every man has his own conception of the world, and this he cannot so easily lay aside. We must, for example, use language, and our language is necessarily steeped in preconceived ideas. Only, they are unconscious preconceived ideas, which are a thousand times the most dangerous of all.

Let's look at this in a bit more detail. Consider his statement: *would it make every experiment fruitless.* I have served on many review panels and refereed many proposals. Not one of them was free of preconceived ideas or was truly open minded. I guess such a proposal would begin: *To be open-minded to all points of view and to avoid preconceived ideas and prejudice we have used a random number generator to choose the beam species and energy.* As I say, I have never seen a proposal like that, but I can easily imagine how it would be treated. Not kindly. Review committees are notoriously closed minded. They demand that every proposal justify the work based on the current understanding in the field. The value of an experiment depends on how it relates to the current models in the area. The experiments at the Large Hadron Collider (LHC) are given meaning by the standard model of particle physics. Every experiment at TRIUMF has to be justified based on what it will tell us about how it fits into the nuclear models.

What about the acceptance of new ideas? Surely, there, we have to be open minded. Certainly not! Extraordinary claims require extraordinary proof. This is not a statement of open mindedness. The idea here goes back at least to Pierre-Simon Laplace (1749 – 1827): *The weight of evidence for an extraordinary claim must be proportioned to its strangeness.* We saw this closed mindedness play out recently with respect to neutrinos traveling faster than the speed of light. The initial claim was roundly rejected; the proponents criticized for publishing such a preposterous idea. In this case, the closed-minded people were correct (they frequently are) as it was subsequently found that there was an experimental error.

Even if we wanted to be, we could not be open minded. Frederick II (1194 – 1250) is said to have carried out an experiment were he had infants raised without people talking to them to see what the natural language was. What he found was that infants treated this way died. Even independent of that experiment, we know most children are talked to and pick up language and other preconceived ideas from their caregivers. As Poincaré said, language is steeped in preconceived ideas. A truly open mind, free from preconceived ideas, is an imspossibility.

Continuing Poincaré's quote:

> *Shall we say, that if we cause others [preconceived ideas] to intervene of which we are fully conscious, that we shall only aggravate the evil? I do not think so. I am inclined to think that they will serve as ample counterpoises—I was almost going to say antidotes. They will generally disagree, they will enter into conflict one with another, and ipso facto, they will force us to look at things under different aspects. This is enough to free us. He is no longer a slave who can choose his master.*

If you like, we should choose our preconceived ideas and choose them wisely. Then we are in charge, not them.

Open-mindedness and freedom from preconceived ideas are only positive in small doses. One has to be open minded enough to accept the next breakthrough, but not so open minded as to follow every will-o'-the-wisp. The real genius in science is in knowing when to be open minded and when to be as stubborn as a mule. It is in knowing which ideas to hold onto and which one to discard.

75. THE MYTH OF THE RATIONAL SCIENTIST

There is this myth that scientists are unemotional, rational seekers of truth. This is typified by the quote from Bertrand Russell: *But if philosophy is to attain truth, it is necessary first and foremost that philosophers should acquire the* **disinterested intellectual curiosity** *which characterises the* **genuine man** *of science* [emphasis added]. But just get any scientist going on his pet theory or project, and any illusion of disinterest will vanish in a flash. *A scientist in his laboratory is not a mere technician: he is also a child confronting natural* *phenomena that impress him as though they were fairy tales[1]*. I guess most scientists are not genuine men, or women, of science. Scientists, at least successful ones, are marked more by obsession than disinterested intellectual curiosity. They are people who wake up at one in the morning and worry about factors of two or missed systematic errors in their experiments, people who convince themselves that their minor role is crucial to the great experiment, people who doggedly pursue a weakly motivated theory or experiment. In the end, most fade into oblivion, but some turn out spectacularly successful and that motivates the rest to keep slugging along. It's a lot like trying to win the lottery.

The obsession leads to a second myth; that of the mad scientist: cold, obsessed to the point of madness, and caring only about his next result. The mad scientist who has both a mistress and a wife so that while the wife thinks he is with the mistress and the mistress thinks he with the wife, he is down at the laboratory getting some work done. The myth is typified by the character Dr. Faustus, who sold his soul to the devil for knowledge, Dr. Frankenstein from Mary Shelley's (1797 –1862) book, or in real life, by the likes of Josef Mengele[2] (1911 – 1979). The mad scientist has also been a staple of movies and science fiction. But most real scientists are not *that* obsessed and all successful people, regardless of their field—science, sports or business—are driven.

In terms of pettiness, Sir Isaac Newton (1642 – 1727) takes the cake. He carefully removed references to Robert Hooke (1635 – 1703) and Gottfried Leibniz (1646 –1716) from versions of the Principia. In Newton's defense, it can be said that the forger, William Chaloner (~1665 – 1699), was the only person he had drawn and quartered. I do not know of modern scientists taking things to that extreme, but there is a recorded case of one distinguished professor hitting another over the head with a teapot. According to the legend, the court ruled it justified. I guess it was the rational and disinterested thing to do. There is also an urban legend of a researcher urinating on his compet-

---

[1] Quoted from Marie Curie (1867 – 1934).
[2] A notorious Nazi physician at Auschwitz.

itor's equipment. The surprising thing is that these reports, even when not true, are at least credible.

In a similar vein, it has been suggested that many great scientists have suffered from autism or Asperger's syndrome. These include Henry Cavendish (1731 – 1810), Charles Darwin (1809 – 1882), Paul Dirac (1902 – 1984), Albert Einstein (1879 – 1955), Isaac Newton (1642 – 1727), Charles Richter (1900 – 1985) and Nikola Tesla (1856 – 1943). Many of these diagnoses have been disputed, but it indicates that some of the symptoms of autism were present in these scientists' behaviour, for example, the single-mindedness with which they pursued their research.

So, are scientists disinterested, autistic, overly obsessed, and/or mad? Probably not more than any other group of people. But to be successful in any field—and especially in science—is demanding. To become a scientist requires a lot of work, dedication, and talent. Consider the years in university. Typically there are four years as an undergraduate. It is at least another four years for a Ph.D. and typically longer. Then to become an academic, you have to spend a few years as a Post-Doctoral Fellow. It is a minimum of ten years of hard work after high school to become an academic. In my case, it was thirteen years from high school to a permanent job. To become a scientist, you have to be driven. Even after you become a scientist, you have to be driven to stay at or near the top. It is not clear if scientists are driven more by a love of their field, or by paranoia. I have seen both and they are not mutually exclusive.

If scientists really were the bastions of rationality that they are sometimes portrayed to be, science would probably grind to a halt. Most successful ideas start out half-baked in some scientist's mind. Only scientists willing to flog such half-baked ideas can become famous[1]. To become successful, an idea must be pursued before there is any convincing evidence to support it. It is only after the work is done that there can be reasons to believe it. Those who succeed in making their ideas mainstream are made into heroes, those that fail, into crackpots. Generally, it is a bit of a crapshoot.

---

[1] Perhaps, that was my mistake.

While individual scientists are not disinterested, nor driven by logic rather than emotion, science as an enterprise is. The error control methods of science, especially peer review and independent repetition, average the biases and foibles of individual scientists to give reliable results. No one should be particularly surprised when results that have not undergone this vetting, particularly the latter, are found to be wrong[1]. However, in the final analysis, the enterprise of science reflects the personality of its ultimate judges: observation and parsimony. They are notoriously hard-hearted, disinterested, and unemotional.

76. IS THE UNDERSTANDABILITY OF THE UNIVERSE A MIRAGE?

Isaac Asimov (1920 – 1992) *expressed a certain gladness at living in a century in which we finally got the basis of the universe straight.* Albert Einstein (1870 – 1955) claimed: *The most incomprehensible thing about the world is that it is comprehensible.* Indeed there is general consensus in science that not only is the universe comprehensible but is it mostly well described by our current models. However, Daniel Kahneman (b. 1943) counters: *Our comforting conviction that the world makes sense rests on a secure foundation: our almost unlimited ability to ignore our ignorance.*

Well, that puts a rather different perspective on Asimov's and Einstein's claims. So who is this person that is raining on our parade? Kahneman is a psychologist who won the 2002 Nobel Prize in economics for his development of prospect theory. A century ago everyone quoted Sigmund Freud (1856 – 1939) to show how modern they were. Today, Kahneman seems to have assumed that role.[2]

Kahneman's Nobel Prize-winning prospect theory, developed with Amos Tversky (1937 – 1996), replaced expected utility theory. The latter assumed that people made economic choices based on the expected utility of the results that is they would behave rationally. In contrast, Kahneman and company have shown that people are irra-

---

[1] Hence, the recently noted medical research results that were wrong.
[2] Let's hope time is kinder to Kahneman than it was to Freud.

tional in well-defined and predictable ways. For example, it is understood that the phrasing of a question can (irrationally) change how people answer although the meaning of the question is the same.

Kahneman's book, THINKING, FAST AND SLOW, really should be required reading for everyone. It explains a lot of what goes on (gives the illusion of comprehension?) and provides practical tips for thinking rationally. For example, when I was on a visit in China, the merchants would hand me a calculator to type in what I would pay for a given item. Their response to the number I typed in was always the same: *You joking, right?* Kahneman would explain that they were trying to remove the anchor set by the first number entered in the calculator. Anchoring is a common aspect of how we think.

Since, as Kahneman argues, we are inherently irrational one has to wonder about the general validity of the philosophic approach to knowledge; an approach based largely on rational argument. Science overcomes our inherent irrationality by constraining our rational arguments by frequent, independently-repeated observations. Similarly in project management, we tend to be irrationally over confident of our ability to estimate resource requirements. Estimates of project resource requirements not constrained by real world observations lead to the project being over budget and delivered past deadlines. Even Kahneman was not immune to this trap of being overly optimistic.

Kahneman's cynicism[1] has been echoed by others, for example H.L. Mencken (1880 –1956) said: *The most common of all follies is to believe passionately in the palpably not true. It is the chief occupation of mankind.* Are the cynics correct? Is our belief the universe is comprehensible, and indeed mostly understood, a mirage based on our unlimited ability to ignore our ignorance? A brief look at history would tend to support that claim. Surely the Buddha (born between the sixth and fourth centuries BCE), after having achieved enlightenment, would have expressed relief and contentment for living in a century in which we finally got the basis of the universe straight. Saint Paul (c. 5 – c. 67), in his letters, echoes the same claim that the

---

[1] Although he claims not to be a cynic.

universe is finally understood. René Descartes (1596 – 1650), with the method laid out in the DISCOURSE ON THE METHOD AND PRINCIPLES OF PHILOSOPHY, would have made the same claim. And so it goes, almost everyone down through history believes that he or she comprehends how the universe works. I wonder if the cow in the barn has the same illusion. Unfortunately, each has a different understanding of what it means to comprehend how the universe works so it is not even possible to compare the relative validity of the different claims. The unconscious mind fits all it knows into a coherent framework that gives the illusion of comprehension in terms of what it considers important. In doing so, it assumes that what you see is all there is. Kahneman refers to this as WYSIATI (What You See Is All There IS).

To a large extent the understandability of the universe is mirage based on WYSIATI—our ignorance of our ignorance. We understand as much as we are aware of and capable of understanding; blissfully ignoring the rest. We do not know how quantum gravity works, if there is intelligent life elsewhere in the universe[1], or for that matter what the weather will be like next week. While our scientific models correctly describe much about the universe, they are in the end only models, and leave much beyond their scope, the ultimate nature of reality for example.

77. WOMEN IN PHYSICS AND MATHEMATICS

Dedicated to Johanna[2]

There are two observations about women in physics and mathematics that are at odds with each other. The first is that there are relatively few women in science. In a typical seminar or conference presentation I have counted that just over ten percent of the audience is female. The second is that, despite the relatively few women, they are by no means second-rate scholars. The first person to ever win two

---

[1] Given our response to global warming, one can debate if there is intelligent life on earth.
[2] A fellow graduate student who died many years ago of breast cancer.

Nobel Prizes was a woman—Marie Curie (1867 – 1924), so did her daughter. But I do not have to go far-far away and long-long ago to find first rate women scientists. I just have to go down the corridor, well actually down the corridor and up a flight of stairs since my office is in the ground floor administrative ghetto while the real work gets done on the second floor. Since women are demonstratively capable, why are there so few of them in the mathematical sciences?

A cynic could say they are too bright to waste their time on such dead end fields[1] but as a physicist I could never admit the validity of that premise. So why are there so few women in physics and mathematics? It is certainly true that in the past these subjects were considered too hard or inappropriate for women. Despite her accomplishments and two Nobel prizes, Madam Curie was never elected to the French Academy of Sciences. Since she was Polish as well as a woman the reason may have been as much due to xenophobia as misogyny.

Another interesting example of a successful woman scientist is Caroline Herschel (1750 – 1848). While not as famous as her brother William (1738 – 1822), she still made important discoveries in astronomy including eight comets and three nebulae. The comment from WIKIPEDIA is in many ways typical: *Caroline was struck with typhus, which stunted her growth and she never grew past four foot three. Due to this deformation, her family assumed that she would never marry and that it was best for her to remain a house servant. Instead she became a significant astronomer in collaboration with William.* Not attractive enough to marry and not wanting to be a servant she made lasting contributions to astronomy. If she had been considered beautiful we would probably never have heard of her! Sad.

Sophie Germain (1776 – 1831) is another interesting example. She overcame family opposition to study mathematics. Not being allowed to attend the lectures of Joseph Lagrange (1736 – 1813) she obtained copies of his lecture notes from other students and submitted assignments under an assumed male name. Lagrange, to his credit, became her mentor when he found out that the outstanding student was a woman. She also used a pseudonym in her correspondence with Carl

---

[1] My wife would probably agree with this.

Gauss[1] (1777 – 1855). After her death, Gauss made the comment: *[Germain] proved to the world that even a woman can accomplish something worthwhile in the most rigorous and abstract of the sciences and for that reason would well have deserved an honorary degree.* High praise from someone like Gauss, but why: *even a woman?* It reminds one of the quote from Voltaire (1694 – 1778) regarding the mathematician Émilie du Châtelet (1706 – 1749): *a great man whose only fault was being a woman.* Fault? And so it goes. Even outstanding women are not allowed to stand on their own merits but denigrated for being women.

But what about today, does this negative perception still continue? While I have observed that roughly ten percent of attendees at physics lectures tend to be female, the distribution is not uniform. There tend to be more women from countries like Italy and France. I once asked a German colleague if she thought Marie Curie as a role model played a role in the larger (or is that less small) number of female physicists from those counties. She said no, that it was more to do with physics not being as prestigious in those counties. Cynical but probably true[2]; through prejudice and convention women are *delegated to* roles of less prestige rather than those reflecting their interests and abilities.

My mother is probably an example of that. The only outlet she had for her mathematical ability was tutoring hers and the neighbour's children, and filling out the family income tax forms. From my vantage point, she was probably as good at mathematics as many of my colleagues. One wonders how far she could have gone given the opportunity, a B. Sc., a Ph. D? One will never know. The social conventions and financial considerations made it impossible. Her sisters became school teachers while she married a small time farmer and raised five children. It is a good thing she did because otherwise I would not exist[3].

---

[1] Probably the greatest mathematician that ever existed.
[2] Recent studies support her contention.
[3] Note the conflict of interest.

78. HAS THERE EVER BEEN A PARADIGM SHIFT?

Yes, once!

Paradigm and *paradigm shift* are so overused and misused that the world would benefit if they were simply banned. Originally Thomas Kuhn (1922 – 1996) in his 1962 book, THE STRUCTURE OF SCIENTIFIC REVOLUTIONS, used the word paradigm to refer to the set of practices that define a scientific discipline at any particular period of time. A paradigm shift is when the entire structure of a field changes, not when someone simply uses a different mathematical formulation. Perhaps it is just grandiosity, everyone thinking their latest idea is earth shaking (or paradigm shifting), but the idea has been so debased that almost any change is called a paradigm shift, down to the level of changing the colour of ones socks.

The archetypal example, and I would suggest the only real example in the natural and physical sciences, is the paradigm shift from Aristotelian to Newtonian physics. This was not just a change in physics from *the perfect motion is circular* to *an object either is at rest or moves at a constant velocity, unless acted upon by an external force* but a change in how knowledge is defined and acquired. There is more here than a different description of motion; the very concept of what is important has changed. In Newtonian physics there is no place for *perfect motion* but only rules to describe how objects actually behave. Newtonian physics was driven by observation. Newton himself went further and claimed his results were derived from observation. While Aristotelian physics is broadly consistent with observation, it is driven more by abstract concepts like perfection. Aristotle (384 BCE – 322 BCE) would most likely have considered Galileo Galilei's (1564 – 1642) careful experiments beneath him. Socrates (c. 469 BCE – 399 BCE) certainly would have. Their epistemology was not based on careful observation.

While there have been major changes in the physical sciences since Newton (1642 – 1726), they do not reach the threshold needed to call them paradigm shifts since they are all within the paradigm defined by the scientific method. I would suggest Kuhn was misled by the

Aristotle-Newton example where, indeed, the two approaches are incommensurate: what constitutes a reasonable explanation is simply different for the two men. But would the same be true with Michael Faraday (1791 – 1867) and Niels Bohr (1885 – 1962) who were chronologically on opposite sides of the quantum mechanics cataclysm? One could easily imagine Faraday, transported in time, having a fruitful discussion with Bohr. While the quantum revolution was indeed cataclysmic, changing mankind's basic understanding of how the universe worked, it was based on the same concept of knowledge as Newtonian physics. You make models based on observations and validate them through testable predictions. The pre-cataclysmic scientists understood the need for change due to failed predictions, even if, like Albert Einstein (1879 – 1955) or Erwin Schrödinger (1887 – 1961), they found quantum mechanics repugnant. The phenomenology was too powerful to ignore.

Sir Karl Popper (1902 – 1994) provided another ingredient missed by Kuhn, the idea that science advances by the bold new hypothesis, not by deducing models from observation. The Bohr model of the atom was a bold hypothesis not a paradigm shift, a bold hypothesis refined by other scientists and tested in the crucible of careful observation. I would also suggest that Kuhn did not understand the role of simplicity in making scientific models unique. It is true that one can always make an old model agree with past observations by making it more complex[1]. This process frequently has the side effect of reducing the old model's ability to make predictions. It is to remedy these problems that a bold new hypothesis is needed. But to be successful, the bold new hypothesis should be simpler than the modified version of the original model and more crucially must make testable predictions that are confirmed by observation. But even then, it is not a paradigm shift; just a verified bold new hypothesis.

Despite the naysaying, Kuhn's ideas did advance the understanding of the scientific method. In particular, it was a good antidote to the logical positivists who wanted to eliminate the role of the model or what Kuhn called the paradigm altogether. Kuhn made the point that

---

[1] This is known as the Duhem-Quine thesis.

it is the framework that gives meaning to observations. Combined with Popper's insights, Kuhn's ideas paved the way for a fairly comprehensive understanding of the scientific method.

But back to the overused word *paradigm*, it would be nice if we could turn back the clock and restrict the term *paradigm shift* to those changes where the before and after are truly incommensurate; where there is no common ground to decide which is better. Or if you like, the demarcation criteria for a paradigm shift is that the before and after are incommensurate[1]. That would rule out the change of sock color from being a paradigm shift. However, we cannot turn back the clock so I will go back to my first suggestion that the word be banned.

79. SELLING SCIENCE

Too many of the attempts to sell science are like the proverbial minister preaching to the choir: they convince nobody but the already converted. This is unfortunate because we, as scientists, have a duty and a responsibility to sell science. There are four motivations for this:

1. There are important technical questions that can only be answered by the scientific method. These include, for example, what is causing global warming? Or why are the returning salmon runs in British Columbia so erratic? We must make the case that science and only science can address these types of questions and that the answers science provides should be listened to.

2. For science to flourish and provide answers to the questions above it must have the support of the general public and politicians on an ongoing basis. Answers to those questions can only come from a scientific infrastructure that is maintained for the long haul. In addition to answering practical questions science has the important cultural role of satisfying human curiosity. For the long term viability of science and its cultural

---

[1] There are probably paradigm shifts, even in the restricted meaning of the word, if we go outside science. The French revolution could be considered a paradigm shift in the relation between the populace and the state.

component we need support from the public purse. This means science must be sold to politicians and the general public.

3. We need to excite the next generation's best and brightest to consider science as a career. This is the only way that we can insure science's future. There is more to life than making oodles of money.

4. It is fun. You should have seen the fun both TRIUMF staff and visitors had at the last TRIUMF open house.

The allusions to religion in the opening sentence are appropriate as many attempts to sell science come across as a claim that science is the one true religion and anyone who disagrees is a fool. While that may, indeed, be true[1], hollering it from the hilltops is a strategy doomed to failure. A frontal attack on a major component of a person's world view will only arouse hostility. Hence, to sell science, we have to start with a common ground with the audience. As with any talk, blog, book etc., you have to know the audience and tailor what you say to that audience. However, there are three things that should be part of any attempt to sell science:

1. A definition of what science is. This may seem self-evident but I have seen seminars on selling science that carefully avoided any attempt to define what science actually is. I have this real nice pig in the poke to sell you. Even worse are attempts to define science that are wrong and/or annoy people. A major impediment to selling science is that there is no commonly accepted definition of what science is. However, a fairly safe definition is: using observation as a basis for modeling how the universe works. Understatement is frequently more effective than overstatement and this definition is simple, understandable and not offensive to most audiences. Or alternatively, one can talk about the ability to make testable predictions as the hallmark of the scientific method. Use the word theory sparingly as that word has multiple meanings and invariably leads to confusion. Using words like objective real-

---

[1] Or not, as the case may be.

207

ity, truth, or fact is a real turn off to many audiences. Besides every Christian will tell you that Jesus is the truth and the more fundamentalist that the Bible is fact. You cannot win those arguments, avoid them.

2. Examples of scientific successes. This is science's greatest strength. We have a plethora of examples to choose from, but it is probably not a good idea to start with the nuclear bomb. Again, it is important to understand the audience. To a person talking non-stop on his cellphone, the cell phone would be a good example (if you can get his attention) but to other people the cellphone is an anathema. The same is true of almost any example you can choose. After all, curing disease (and motherhood) leads to world overpopulation. On TV or radio, the role of science in enabling TV and radio is a good bet. On YouTube, the internet would be a good example. Despite the comment above, curing disease usually gets brownie points for science. But claiming the Higgs boson cures cancer is a bit of a stretch.

3. Your personal experience of the thrill of science; whether it is for the good of humanity or just learning more about how the universe works. It is here that the emotional aspect of science can come to the fore. To some of us, the hunting of the Higgs boson is more thrilling than hunting grizzly bears and probably more environmentally friendly. Using personal experience may seem like going against our training as scientists; but here we can learn from the professionals, those who sell religion or political parties: Do not talk about theology but your personal experience[1]. Do not talk about the platform but your own experience[2]. In the end, this may be a telling argument and it is important to counter the stereotype of the mad scientists (almost always male) in his laboratory plotting world domination or ignoring the obvious flaws in his theory and its disastrous side effects. Drs. Faustus and Frankenstein are never far from people's conception of the scientist.

---

[1] A well-known mega church pastor.
[2] An Obama campaign worker.

You would think that selling science would be easy. We have a well-defined technique, four hundred years of successes to prove its usefulness and the thrill of the hunt. But we are up against two formidable foes: competing world views and vested interests. If someone believes they will be raptured to Heaven in the near future, learning about this world below is not a high priority. Similarly if they subscribe to the old hymn, *I Don't Want to Get Adjusted to This World Below*[1], finding a crack to start a conversation is difficult.

In the same vein, if you have spent your life building a tobacco empire the last thing you want is some scientist claiming tobacco causes cancer. Or if you have made selling tar sands oil a key political plank, you do not want scientists claiming it is destroying the earth. In these cases, science itself tends to become the target of the counterattack. With the world's best public-relations machines powered by religion and vested interests in opposition it is not at all clear the efforts to sell science will be successful. But we must try, we must try.

80. SCIENCE AND THE ARTS: COMPLEMENTARY PARADIGMS

Man does not live by science alone.

This essay is motivated in part[2] by an excellent Quantum Diary post, ART AND SCIENCE: BOTH OR NEITHER, written by a fellow TRIUM-Fonian, Jordan Pitcher. He explores the incommensurability of art and science. In response, this essay is about related matters of art and taste such as if the original Bugs Bunny cartoons represent the pinnacle of the cartoon trade (beyond debate) or if my essays are worth reading (debatable).

Now, science is about learning how the universe operates, but there is more to life than that; he says while listening to: *Oh, give me the beat boys and free my soul, I wanna get lost in your rock and roll, And*

---

[1] By Sanford J. Massengale (1929 – 1992).
[2] And in part by my wife's interest in literature.

drift away[1]. I do not know if that is art but it is certainly not science. This brings us to the aesthetic side of life, including music, art, literature, theatre, movies, and cartoons (see above). With art, my house is full of the pictures my mother-in-law painted and it ain't half bad stuff. I particularly like the autumn scenes. Similarly, I have my tastes in literature (I particularly like a good Rex Stout (1886 – 1975) detective story—and don't say that is not literature); theatre (WICKED was not bad); movies (I took my daughter to THE SMURFS [2] but can skip movies with no loss); and the appropriate use of grammar (ain't, ain't half bad[3]). That brings me to fonts: Is **comic sans** really that bad? And I have seen major disagreements on the relative merits of *sans serif* versus *serif*—a plague on both your houses. And do not forget food; a fried mackerel would go real good about now.

I lived through the culture wars of the 1960s and it turned me off culture wars—meaningless arguments over personal preferences. There were great debates about whether rock and roll was legitimate music or the work of the devil. There was even a debate about whether the lyrics of one particular song (LOUIE LOUIE, a 1963 hit by The Kimgsmen) were obscene, but no one could tell what the lyrics actually were so the argument was moot. Similarly, is abstract art really art or can art only be more realistic works like Rubens (1557 – 1640)? Hmm, perhaps that is not the best example but there can be no doubt that the arts enrich life. Again, perhaps Rubens might not be the best example.

A central characteristic of science is that it has mechanisms, comparison to observation, and parsimony to uniquely determine which model or approach is best. But there are no similar criteria to decide if Beethoven (1770 – 1827) is really better than the Beastie Boys (1981 – 2012). I do not particularly like either (*Oh, give me the beat*). Considering Beethoven's staying power and the Beastie Boys record sales, I guess I am in a minority. So it is across the whole scope of the arts;

---

[1] From the song DRIFT AWAY written by Mentor Williams.
[2] I particularly liked Azrael although my sister says I am like brainy smurf (not a compliment).
[3] See A DICTIONARY OF MODERN ENGLISH USAGE (1926), by Henry Fowler (1858 – 1933).

some people like one thing and some another. One should not mistake personal preference for objective reality, but *give me that beat.*

It is also not a scientist versus humanitarian kind of thing. The likes and dislikes cut across that divide. Perhaps the likes of scientists may be tilted in a somewhat different direction but the spread in each group is large. There is, instead, a large upbringing and cultural influence on what ones likes and dislikes. My Asian born daughter loves sushi but my Nova Scotian raised relatives would not touch it with a three metre (roughly ten foot) pole. The choice of preferred music, art, food, etc. depends at least in part on what one was exposed to while growing up. There is probably even a genetic component to what one likes and dislikes. I inherited my like for and ability in mathematics from my mother; similarly my inability to write coherently. The latter plagued my time in school and university. What one sees has a genetic component; as in colour blindness. There are also studies suggesting some women have four rather than three types of colour sensors. It would be strange if inherited differences such as these did not affect our aesthetic tastes. Indubitably, some of the differences in our tastes are indeed in our brains—either acquired or inherited.

The one downside of all the differences in taste is that some denizens of the art world think that since the arts have no or only weak objective standards, science cannot have any either. This leads to nonsense like the claim that science is purely cultural. Conversely, there is the equally ridiculous perception that the arts should have objective standards like science. Salt herring is an acquired taste (shudder).

So let us recognize that science and the arts are indeed very different in how they make judgments and celebrate the diversity permitted by the subjectivity in the arts. After all, life would be very boring if all we had to read was Margaret Atwood (b. 1939) or Farley Mowat, (b. 1921 – 2014).

81. SCIENCE AND PHILOSOPHY: COMPETING PARADIGMS

For the antepenultimate[1] essay in this series, I will tackle the thorny issue of the relation between science and philosophy. Philosophy can be made as wide as you like to include anything concerned with knowledge. In that regard, science could be considered a subset of philosophy. It is even claimed that science arose out of philosophy, but that is an oversimplification. Science owes at least as much to alchemy as to Aristotle. After all, both Isaac Newton (1642 – 1727) and Robert Boyle[2] (1627 – 1691) were alchemists and the philosophers, including Francis Bacon (1561 – 1626), vehemently opposed Galileo (1564 – 1642). Here, I wish to restrict philosophy to what might be call western philosophy—the tradition started with the ancient Greeks and continued ever since in monasteries and the hallowed halls of academia.

Let us start this discussion with Thomas Kuhn (1922 – 1996). He observed that Aristotelian physics and Newtonian physics did not just differ in degree, but were entirely different beasts. He then introduced the idea of paradigms to denote such changes of perspective. However, Kuhn misidentified the fault line. It was not between Aristotelian physics and Newtonian physics, but rather between western philosophy and science. Indeed, I would say that science—along with its sister discipline, engineering—is demarcated by a common definition of what knowledge is (see below). In science, classical and quantum mechanics are very different, yet they share a common paradigm for the nature of knowledge and, hence, we can compare the two from common ground.

Bertrand Russell (1872 –1970) in his A HISTORY OF WESTERN PHILOSOPHY makes a point similar to Kuhn's. Russell claims that from the ancient Greeks up to the renaissance, philosophers would have been able to understand and discourse with each other. Plato (424 BCE – 348 BCE) and Machiavelli (1469 –1527) would have been

---

[1] That is N^{2}LP in the compact notation of effective field theorists.
[2] The son of the Earl of Cork and the father of modern chemistry.

able to discuss, if brought together. Similarly with Thomas Aquinas (1225 – 1274) and Martin Luther (1483 – 1546), if Aquinas refrained from having Luther burnt at the stake. They shared a common paradigm, if not a common view. But with the advent of science, that changes. Neither Aristotle nor Aquinas would have understood Newton. The paradigm had shifted. This shift from philosophy to science is the best and, perhaps, the only real example of a paradigm shift in Kuhn's original meaning. Like Kuhn, Russell misidentified the fault line. It was not between early and late western philosophy, but between philosophy and science. C.P. Snow (1905 – 1980) in his 1959 lecture, THE TWO CULTURES, identifies a similar fault line but between science and the humanities more generally.

So what are these two paradigms? Philosophy is concerned with using rational arguments[1] to understand the nature of reality. Science turns that on its head and defines rational arguments through observation. A rotational argument is one that helps build models that can be used to predict the future. To doubt that Euclidian geometry describes physical space-time or to suggest twins could age at different rates were at one time considered irrational ideas, beyond the pale. But now they are accepted due to observation-based modeling. Philosophy tends to define knowledge as that which is true and known to be true for good reason (with debate over what good reason is). Science defines knowledge in terms of observation and observationally constrained models with no explicit mention of the metaphysical concept of truth. Science is concerned with serviceable knowledge, rather than certain knowledge[2].

Once one realizes science and philosophy are distinct paradigms, a lot becomes clear. For example, why philosophers have had so much trouble coming to grips with what science is. Scientific induction as proposed by Francis Bacon (1561 – 1626) does not exist. David Hume (1711 – 1776) started the philosophy of science down the dead-end street to logical positivism. Immanuel Kant (1724 – 1804) thought Euclidean geometry was synthetic *a priori* information, and Karl Popper (1902 – 1994) introduced falsification, which is now

---

[1] This is an oversimplification but sufficient for our purposes.
[2] An idea picked up by the pragmatic philosophers.

largely dismissed by philosophers. Even today, the philosophic community as a whole does not understand what the scientific method is and tends toward the idea that it does not exist at all. All attempts, by either scientist or philosophers, to fit the square peg of science into the round hole of western philosophy have failed and will probably continue to do so into the indefinite future. Eastern philosophy is even more distant.

The different paradigms also provide the explanation of the misunderstanding between science and philosophy. Alfred Whitehead (1861 – 1947) claimed that all of modern philosophy is but footnotes to Plato. On the other hand, Carl Sagan (1934 – 1996) claims Plato and his followers delayed the advance of knowledge by two millennia. The two statements are not in contradiction if you have a negative conception of philosophy. And indeed, many scientists do have a negative conception of philosophy; a short list includes Richard Feynman (1918 – 1988), Ernest Rutherford (1871 – 1937), Steven Weinberg (b. 1933), Stephen Hawking (b. 1962), and Lawrence Krauss (b. 1954). Feynman is quoted as saying: *Philosophy of science is about as useful to scientists as ornithology is to birds.* To a large extent, Feynman is correct. The philosophy of science has had little or no effect on the actual practice of science. It has, however, had a large impact on the scientist's self-image of what they do. Newton was influenced by Francis Bacon, Darwin by Hume, and just try suggesting to a room full of physicists that science is not based on falsification[1]. Even this essay is built around Kuhn's concept of a paradigm (but most of Kuhn's other ideas on science are, to put it bluntly, wrong).

This series of essays has been devoted to defining the scientific paradigm for what knowledge is. The conclusion I have reached, as noted above, is that western philosophy (excluding parts of the pragmatic tradition) and science are based on different paradigms for the nature of knowledge. But are they competing or complementary paradigms? My take is that the two paradigms are incompatible as well as in-

---

[1] Although I am a theorist, I did that experiment. Not pretty.

commensurate. Knowledge cannot be simultaneously defined by what is true in the metaphysical sense, and by model building.

82. SCIENCE AND THEOLOGY: COMPETING PARADIGMS

The contentious relation between science and religion is the topic of this, the penultimate[1] essay in this book. Ever since science has gone mainstream, there have been futile attempts to erect a firewall between science and religion. Galileo got in trouble with the Catholic Church, not so much for saying the earth moved as for suggesting the church steer clear of scientific controversies[2]. More recently, we have methodological naturalism (discussed in a previous essay), a misidentification of why the supernatural is absent from science. Then there is the: *science cannot answer the why question*—but it can when it helps make better models (also discussed in a previous essay). For example, why do beavers build dams? This can be answered by science. And there is the ever-popular non-overlapping magisteria (NOMA) of Stephen J. Gould (1941 – 2002). NOMA claims that *the magisterium of science covers the empirical realm: ... The magisterium of religion extends over questions of ultimate meaning and moral value.*

The empirical realm covers not just what can be directly observed but what can be implied from what is observed. For example, quarks, and even something as well-known as electrons, are not directly observed but are implied to exist. That would also be true for citizens of the spirit or netherworld. If they exist, they presumably have observable effects. If they have no observable effect, does it matter if they exist or not? Similarly, a religion with no empirical content would be quite sterile, i.e. would prayer be meaningful if it had absolutely no observable effects?

---

[1] NLP in the notation of effective field theorists.
[2] Stillman Drake (1910 – 1993), GALLILEO: A VERY SHORT INTRODUCTION (1980).

Moral issues cannot be assigned purely to the religious sphere. The study of brain function impacts questions of free will and moral responsibility. Disease and brain injury can have very specific effects on behaviour, for example, a brain injury led to excessive swearing in one person. What about homosexuality? Is it biological or a lifestyle choice? Recent research has indicated a genetic component in homosexuality, thus mixing science with what some regard as a moral issue. Finally, what about when life begins and ends? Who decides who is dead and who is alive? And by what criteria? Scientific or religious? This has huge implications for when to remove life support. The bigger fight is over abortion and the question of when independent life begins. Is it when the sperm fertilizes the egg? That is a scientific concept developed with the use of the microscope. That simple definition has problems when there are identical twins. There the proto-fetus splits in two much later than at conception. In the other direction, both the sperm and the egg can be considered independent life. After all, the sperm has the ability to leave the donor's body and survive for a period of time. The arguments one hears regarding when independent life begins are frequently an ungodly combination of scientific and theological arguments.

In the end, there is only one reality; however we choose to study or approach it. Thus, any attempt to put a firewall between different approaches to reality will ultimately fail, be they based on science, religion, or philosophy. At least the various religious fundamentalists recognize this, but their solution would take us back to the dark ages by subjugating science to particular religious dogmas. However, it does not follow that religion and science have to be in conflict. Since there is so much variation in religions, some are and some are not in conflict with any particular model developed by science. Still, it should be a major concern for theology that something like God has not arisen naturally from scientific investigations. While there are places God can hide in the models science produces, there is no place where He is made manifest. And it is not because He is excluded by fiat either (see the essay on methodological naturalism).

One should not make the same mistake as Andrew Dickson White (1832 –1918) in setting science and religion in perpetual hostility. He was a co-founder of Cornell University and its first president. He was also embittered by the opposition from the church to the establishment of Cornell as a secular institute. The result was the book: HISTORY OF THE WARFARE OF SCIENCE WITH THEOLOGY IN CHRISTENDOM (1896); a polemic against Christianity masquerading as a scholarly publication. This book, along with HISTORY OF THE CONFLICT BETWEEN RELIGION AND SCIENCE by John William Draper (1811 – 1882), introduced the conflict thesis regarding the relation between science and religion and said it is perpetual hostility. Against that, we note Newton, Galileo, and Kepler were all very religious and much science was done by clergymen in nineteenth century England. White's book, in particular, has many problems. One is that the every opposition to change is cast as science versus religion rather than recognizing a lot of it as simple resistance to change. Even science is not immune to that—witness the fifty year delay in the acceptance of continental drift. The historical interplay between science and religion is now recognized to be very complex with them sometimes in conflict, sometimes in concord, and most commonly, indifferent.

If we take a step back from the results of science and its relation to particular religious dogmas, and look instead at the relation between the scientific method and theology, we see a different picture. Like science and western philosophy, science and theology represent competing paradigms for the nature of knowledge. Science is based on observation and observationally constrained models; Western philosophy on rational arguments; while theology is based more on spirituality, divine revelation, and spiritual insight. This is, in many ways, a more serious conflict than between scientific results and particular religions. Particular religions can change, and frequently have changed, in response to new scientific orthodoxy, but it is much more difficult to change one's conceptual framework or paradigm.

As Thomas Kuhn (1922 – 1996) and Paul Feyerabend (1924 – 1994) pointed out, different paradigms are incommensurate, providing different frameworks that make communication difficult. They also have conflicting methods for deciding questions, making the cross-paradigm resolution of issues difficult, if not impossible. For this reason, the prediction, inherited from the Enlightenment, that science will lead to the end of religion has repeatedly failed[1] and will continue to fail for the foreseeable future. Hence, there will be tension between science and theology forever, with neither dominating.

83. SCIENCE: MANKIND'S GREATEST ACHIEVEMENT!

How is that for the ultimate claim in the ultimate[2] essay in this book? Science: mankind's greatest achievement. Can there be any doubt? In the four hundred years since science went mainstream, we have learned how the universe works, changed our conception of man's place in it, and provided the knowledge to develop fantastic technology. We have big history: the inspiring story of the universe beginning with the primordial big bang and creating order out of chaos through self-interaction, and finally life arising and evolving in our corner of the universe. We have developed models that describe the universe on the largest visible scales down to sub-atomic sizes: astronomy, biology, chemistry, cosmology, medicine, physics, psychology, animate, inanimate, eater, and eatee. The models form a mosaic that overlap and interlock to form a seamless whole. An amazingly complete picture. There is still much to know, but let us take credit as scientists, that much is known. And yes, we should be glad to be living in a time when so much is known.

However, science has two shortcomings[3]: it does not offer the illusions of certainty or purpose. I once came across a last will and testament that began: *I commit my body into the ground in the sure and certain knowledge it will be restored to me on the judgment day.* Ah, for sure and certain knowledge. Well, the judgment day has not come yet so we do not know if his sure and certain knowledge was valid, but the resurrection of the body is much less prominent in Christian

---

[1] For a possibly real correlation see: The Chronic Dependence of Popular Religiosity upon Dysfunctional Psychosociological Conditions, Gregory Paul, EVOLUTIONARY PSYCHOLOGY, 2009. 7(3): 398-441.

[2] That is the LP in the language of effective field theorists (LP=last post, not long playing as you old timers thought).

[3] Humility is not one of them.

apologetics than it used to be. When it comes to knowledge, science promises less but delivers more than its competitors in philosophy or theology. I would take Isaac Newton (1642 – 1727) over Rene Descartes (1596 – 1650), Immanuel Kant (1724 – 1804), Thomas Aquinas (1225 – 1274), or William Paley (1743 – 1805) any day of the week and all together. Their certain knowledge has largely vanished, but Newton's uncertain and approximate knowledge is still being used in many practical applications. Ask any mechanical engineer.

In THE HITCHHIKER'S GUIDE TO THE GALAXY, Douglas Adams (1952 – 2001) introduces the total perspective vortex. It was created by a husband whose wife keeps telling him to put things in perspective. However, when anyone looked in the vortex, they realized how utterly insignificant they were in the vast stretches of the universe and invariably went insane and died. This proved that if life is going to exist in a Universe of this size, then the one thing it cannot afford to have is a sense of proportion. Ah yes, the human need for importance and purpose. I guess the best science can come up with for a purpose is entropy[1] generation. I am not sure that is any worse than what I had heard from a Christian apologist who claimed we were created by God to worship him. Personally, I would never worship that narcissistic a God.

Despite its shortcomings, perceived or real, science has a tremendous track record. But the best is still to come. Let us not make the mistake of the late nineteenth century physicists who thought all the important questions had been answered. There are things that enquiring minds still want to know: What, if anything, was there before the big bang? How do you combine gravity and quantum mechanics? Is there a solution for global warming that is politically acceptable? Are there room temperature superconductors? How did life begin? How intelligent were the Neanderthals? How does the mind work? The last strikes me as the most interesting question: the final frontier[2]. It has the potential to open up a whole new front in the conflict between science and religion, or science and philosophy. But it is interesting nonetheless. Answering these questions and others will take clever

---

[1] Entropy generation is the driving force behind evolution.
[2] Sorry Star Trek fans, it is the mind, not space.

theoretical approaches, clever experiments, and clever approaches to funding. However, the techniques of science are up to the task.

But what is science? In the final analysis, it is a human activity, an exercise of the human mind. We construct models and paradigms because that is how our minds and brains have evolved to deal with the complexities of our experiences. Thus, the nature of science is tied closely to the last question asked above: How does the mind work? Ultimately, how science works and indeed, the very definition of knowledge, are questions for neuroscience and the empirical study of the mind.

c. 2650 BCE: birth of Imhotep who helped build the early pyramids and is the earliest known engineer.

c. 490 BCE: birth of Protagoras, an early sophist.

384 BCE: birth of Aristotle, an early philosopher whose ideas misled people for two millennia.

c. 90: birth of Claudius Ptolemy, the creator of the useful, but much maligned Ptolemaic system of planetary motion. It was used for over 500 years.

721: birth of Jābir ibn Hayyān, an Arab scholar, sometimes regarded as the founder of modern chemistry and an early scientist.

1267: publication of OPUS MAJUS by Roger Bacon, an early advocate of the scientific method.

1517: Martin Luther nails his Ninety-Five Theses to the church door and started the Protestant Reformation.

1609: the year Galileo Galilei first turns his telescope on the Heavens, and Johannes Kepler publishes ASTRONOMIA NOVA on elliptic planetary motion.

1687: Isaac Newton publishes the PRINCIPIA MATHEMATICA which established classical or Newtonian mechanics. The ideas presented there are still widely used today.

1781: Immanuel Kant publishes the CRITIQUE OF PURE REASON. Newton and Kant, respectively, roughly mark the beginning and end of the Enlightenment.

1844: Robert Chambers anonymously publishes VESTIGES OF THE NATURAL HISTORY OF CREATION. It laid the groundwork for the acceptance of the theory of evolution. VESTIGES, as it was frequently called, was a best seller and outsold ON THE ORIGIN OF SPECIES (published in 1859 by Charles Darwin) until the early 20th century.

1905: Albert Einstein publishes four papers that laid the foundation for modern physics.

1927: Werner Heisenberg proposes his uncertainty principle. That principle has contributed to the mystique of quantum mechanics ever since.

1951: Willard Quine publishes TWO DOGMAS OF EMPIRICISM, a high point in the philosophy of science.

1953: Francis Crick and James D. Watson publish their work on the structure of DNA. They used results from Rosalind Franklin.

2011: Daniel Kahneman publishes THINKING, FAST AND SLOW, a must read for anyone who wants to understand how we think.

2012: The Higgs boson is discovered at CERN. This confirmed the validity of the Standard Model of particle physics.